NEVER MIND THE B#LL*CKS,
HERE'S THE SCIENCE

WIN THE ADMIRATION
OF YOUR FRIENDS!

NEVER MIND THE B#LL*CKS, HERE'S THE SCIENCE

—

A SCIENTIST'S GUIDE to
the BIGGEST CHALLENGES facing
our SPECIES TODAY

—

PROFESSOR LUKE O'NEILL

GILL BOOKS
Hume Avenue
Park West
Dublin 12
www.gillbooks.ie
Gill Books is an imprint of M.H. Gill and Co.
© Luke O'Neill 2020
978 07171 86396
Designed by www.grahamthew.com
Edited by Djinn von Noorden
Proofread by Neil Burkey
Illustrations by Derry Dillon

'Bed Blocker Blues' from *The Luckiest Guy Alive* by John Cooper Clarke reproduced with permission of the Licensor through PLSclear.

For permission to reproduce photographs, the author and publisher gratefully acknowledge the following:
© Alamy: 4, 13, 30, 32, 34, 43b, 65, 67, 82, 94, 99, 101, 143, 150, 158, 179, 198, 231, 239, 241, 252, 267, 285, 289, 294, 301, 315, 317; © Dara Mac Dónaill/The Irish Times: 249; © Dr Allan Warner via The Jenner Trust: 36; © Fennell Photography: 255; Image courtesy of Chisholm Lab/ Flickr: 272; Image courtesy of F. Opperdoes/Flickr: 296; Image Courtesy of Len Edgerly/Flickr: 257; Image courtesy of LSE Archives/ Flickr: 215; Image Courtesy of Smithsonian's Freer and Sackler Galleries/ Flickr: 124; Image Courtesy of Woodlouse/Flickr: 205; © Aurora Samperio/NurPhoto via Getty Images: 23; © DEA / G. COSTA via Getty Images: 162; © Michael Ochs Archives via Getty Images: 89; © ullstein bild via Getty Images :122; © Universal History Archive via Getty Images: 166; © iStock/ Getty Premium: ii, 2, 8, 16, 21, 26, 50, 56, 70, 90, 97, 112, 116, 118, 120, 128, 134, 138, 147, 154, 176, 180, 188, 200, 221, 237, 244, 262, 282, 298, 305; © NASA: 281; © Shutterstock: 53, 60, 109, 131, 172, 187, 197, 213, 219, 228, 253, 269, 280; © Simon Fraser University: 130; Image courtesy of Southside Travelers: 216; © Trinity College Dublin: 313; © WikiCommons: 12, 18, 39, 77, 105, 137, 141, 161, 208, 211, 226, 248; Image courtesy of Cago Collective: 140; Image courtesy of Dr Tony Ayling/ WikiCommons: 182; Image courtesy of Gage Skidmore/ WikiCommons: 43t; Image courtesy of Guido van Nispen/ WikiCommons: 234; Image courtesy of Nord and USNORTHCOM Public Affairs: 306.

Printed by BZ Graf, Poland
This book is typeset in Freight Text pro Book, 10.7 on 16 pt.
The paper used in this book comes from the wood pulp of managed forests. For every tree felled, at least one tree is planted, thereby renewing natural resources.
A CIP catalogue record for this book is available from the British Library.
5 4 3 2 1

For my sister Helen, who has spent
her life caring for all the lonely people.

ACKNOWLEDGEMENTS

Thanks to Sarah Liddy of Gill Books, who asked me to consider writing a book about how science can inform big questions (but who, I imagine, never thought that 'b#ll*cks' would be in the title ...). Thanks, as ever, for your support, Sarah. Thanks also to Aoibheann Molumby, editor at Gill Books, for excellent editing and many insightful comments.

Several people read the text for me to check facts (and took glee in correcting me) and made some great suggestions: first, Andy Gearing, a fellow immunologist. I met Andy when I was working in the UK and we began a scientific collaboration on an immune cell type he discovered called NOB cells (I won't go into it). More important, one day I went to his flat for lunch and he said, 'See what you can find in the fridge.' All I could find there was a bottle of champagne and a jar of lime pickle. I knew I'd found a lifelong friend. Andy read the whole book twice and came up with so many great suggestions that he should be a co-author. Tough luck, Andy – no royalties for you.

The following people read various chapters: my sister Helen and my wife Margaret both made excellent suggestions for the chapter on men versus women. Cliona O'Farrelly (another fellow immunologist) was a great sounding board for some of the topics in this book. Zbigniew Zaslona (post-doctoral scientist in my lab) made several suggestions and was always great to bounce things off. Brian McManus (who apparently was once a GP, although I find that hard to believe) suggested I include material from *The Matrix* and *The Hitchhiker's Guide to the Galaxy*, as well as suggesting that the list of occupations in the bullshit jobs chapter should include only academics. Brian also made important suggestions for the chapters on euthanasia and racism. Duncan Levy (climate engineer) checked the climate change chapter. My colleague and fellow immunologist Kingston Mills made great suggestions for the vaccines chapter. Aongus Buckley (economist and Thomas Paine fan) and Neil Towart (lefty Australian) made suggestions on the bullshit jobs chapter. Aongus also checked the chapters on racism and on control in life, as did fellow 'Bray-ite' Frances Gleeson (who has more degrees than me). Ken Mealy (surgeon) and Colm O'Donnell (physician) both made suggestions for the chapter on euthanasia. Colm also read the chapters on drug legalisation (he knows a lot about drugs …) and on racism and made important suggestions. Donal O'Shea (physician) made great comments on the dieting chapter and suggested that I include 'fat-shaming'. Chris McCormack (Trim gentleman and former prison governor) made suggestions for the chapters on jail and drug legalisation. And my old mate John O'Connor (neuroscientist) made suggestions for the chapters on addiction and depression. Finally, a big thanks to Stevie O'Neill and Sam O'Neill, who didn't read this book at all, but thanks anyway, lads.

CONTENTS

INTRODUCTION

—

Nothing in life is to be feared,
it is only to be understood.
Now is the time to understand more,
so that we may fear less.

Maria Skłodowska Curie

—

W ELCOME TO *Never Mind The B#ll*cks, Here's The Science.* The title captures exactly what this book is about. It's about the science behind the biggest issues that confront our species today. These issues intrigue me, and hopefully you: control over your life; vaccination; dieting; mental health; addiction; legalising drugs; racism; men and women; bullshit jobs; climate change; euthanasia; the future. Using my scientific training, I've examined the science in each of these topics. I have, in the inimitable words of Matt Damon in the *The Martian*, literally 'scienced the shit' out of them. B#ll*cks and shit? There's two rude words already, and in a science book!

Science is so great because it's based on information that comes from experiments and data that have been independently checked and ultimately reproduced by different scientists. Scientists are competitive and love scientific combat and, when they work together, they are unbeatable. The best want the truth. Science is the antidote to fake news, so we need it now more than ever. We are all astounded by what's happened with COVID-19,

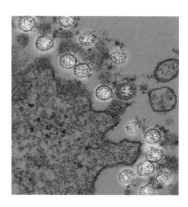

A SAMPLE TAKEN FROM THE FIRST PATIENT INFECTED WITH SARS-COV-2 IN THE US. THE YELLOW SPHERES ARE EACH A SINGLE VIRUS.

the disease caused by the virus SARS-CoV-2. That pandemic has revealed how we need science more than ever, and I discuss several of the topics through the lens of that particular malicious virus. My goal is to get as close to the truth as I can on all of these topics, using science as my only guide.

To be a scientist is to be a sceptic. This is why they argue with each other. I read a great quote recently from a 1924 issue of the Massachusetts Institute of Technology's *Tech Engineering News* magazine (that's how much of a sceptic I am – checking back issues of MIT newsletters). It said, 'The predominant feature of the scientifically trained mind is the ability to associate cause with effect. It is not content with the knowledge and application of facts, but it seeks the reason therefore. It is continually in a state of unrest, turmoil, irritation. Moreover, it manifests a natural inclination and willingness to know the reason "why".' Unrest, turmoil and irritation? Sounds great, doesn't it? But it makes an important point. Scientists obsess about associating cause with effect. Or to put it another way, determining the link between correlation and causation. Vaccines correlating with, but being responsible for protection against, an illness. A diet correlating with, and hopefully causing, weight loss. An antidepressant correlating with, but being the reason for, improved mental health. A genetic variant correlating with, and causing criminal activity. Being female correlating with, and being the reason for, increased empathy. Human activity correlating with, and causing, global warming.

The correlation/causation issue is critical in science: you might well correlate one thing with something else, but that doesn't mean that what is being correlated is the cause of something else. For example, there is a correlation between smoking and cancer. For a long time, tobacco companies claimed that this was just a correlation. But then the science became irrefutable – smoking causes cancer. This can be proven with rigorous statistics and by coming up with a mechanism, which proves the link. In the case of smoking, the mechanism is chemicals in the cigarette smoke: these cause mutations in the genes that produce proteins which, when mutated, cause cancer. The case that smoking causes cancer took off when in 1953 scientists showed that mice painted with tobacco tar developed tumours. Study after study confirmed this observation and extended it, providing the mechanistic link. Game over. This led to panic at tobacco companies who met secretly in New York's Plaza Hotel to start their campaign to counter the bad publicity, dubbed 'the most astonishing corporate deceit of all time'.[1]

Another example of how correlation and causation can become entangled is a study which showed a correlation between the number of babies being born and numbers of storks nesting nearby.[2] This was done to illustrate how we shouldn't jump to conclusions. The investigators found that there was a correlation, and the correlation passed a stringent statistical test. This indicated that storks deliver babies, right? Not so fast. Close inspection revealed that the reason for the correlation was that storks were nesting near larger villages (which had more chimneys for them to build nests on), where the number of newborn babies was greater. So even though there was a correlation, it wasn't occurring because the storks were delivering the babies. The correlation was actually with village size. The bigger the village, the more chimneys; the more chimneys, the more storks but also the more babies being born. The final proof of the claim that storks deliver babies would have been a mechanism, which means showing how it would work: stork scientists observing storks actually delivering babies. You never know – and, as a scientist, I must remain open-minded (and slightly crazy) to even countenance the possibility. And the most important scientists were often crazy because they could think laterally. There's even a study that correlated the places where mad cow disease occurred in the UK with where Brexit voters lived.[3] This one was satirical. We scientists laughed and laughed. But for a split second, we all wondered ...

People don't really want to know that scientists are sceptical. What they want is for scientists to nail their colours to the mast, based on the evidence at hand. Anecdote and snap decisions are the enemy of science – what is needed are experiments, data, statistics and a considered response. Sadly, these criteria can be inconsistent with what politicians or media editors and sometimes even editors of scientific journals want, which is where the problems can begin. Donald Trump's championing of the drug hydroxychloroquine for COVID-19 gives us another egregious example of what can happen when politics and science meet. For political reasons, Trump wanted a rapid treatment for COVID-19. He said the evidence behind hydroxychloroquine as a treatment for COVID-19 was 'great' and 'powerful'. He also asked, 'What do you have to lose?' The president of the American Medical Association, Dr Patricia Harris, answered, 'Your life'. Hydroxychloroquine can damage the heart, and had never been tested against COVID-19, a disease that can involve heart damage. This made doctors very cautious in its use against COVID-19, although it can be used safely for diseases such as rheumatoid arthritis. Dr Anthony Fauci, eminent immunologist and lead member of the Trump administration's White House Coronavirus Task Force, said, 'The data are really just, at best suggestive. In terms of science, I don't think we could definitively say it works.' Who do you agree with? If you're a scientist only one thing matters: data must trump politics.

Deliberation is key to science: it helps us see where the truth lies, as opposed to responding with our intuition. Susceptibility to fake news is due to lazy thinking (you can just hear your old teacher saying that, can't you?) as opposed to inherent bias.

Most of the topics I cover are serious. If we make a wrong decision, we end up destroying the earth. Or killing people on a clinical trial. How does science inform our decisions when it comes to these issues? Each chapter is a question that I address with data and/or experiments to help us (science loves both of these). I've done my best to give you the best evidence to support conclusions drawn. You might want to check the facts yourself, and that's fine. I may have gotten some things wrong. You should correct me – with evidence. That's how it's supposed to work. Now, more than ever, people need to believe in science and scientists, as opposed to what you might read on the side of a bus.

At the start of each chapter, I include a quote to inform the topic from artists, writers or comedians that I love. At the end of each chapter, I give my own bottom line. You could be lazy and just read these, because they capture the essence of each chapter. Or you could do the right thing and delve deeper to find out the science that led me to my conclusions.

I hope the book helps and informs you in your deliberations on these weighty matters. I hope you come away positive about your own life if the topics are especially relevant to you. I hope you will feel more comfortable about these topics: not wholly, of course, because we are built to be uneasy. It's part of our innate nature. Hopefully, you will achieve enlightenment through science. It's only through dialogue and challenges that we make progress. And I hope you feel more positive about the future – a future where we are all headed, with science as our one true friend.

So, with your glass half full, join me on this mission to reaffirm your vows as a scientist if you are one already, or if you've lapsed (no shame in that) to remind you why science is great. And if you're neither of these, I welcome you most warmly. Together we can navigate our way through enormously important questions. And you will see why we should all be scientists: always questioning, always trying to work things out, but ultimately turning darkness into light.

CHAPTER 1

WHAT MAKES YOU THINK YOU'VE CONTROL OVER YOUR LIFE?

—

'We operate on autopilot and end up,
whatever, with a house and family and job and
everything else, and we haven't really stopped
to ask ourselves, "How did I get here?"'

—

David Byrne on his song 'Once in a Lifetime'

YOU PROBABLY THINK you're reading this book because you chose to. You probably think you're free. You move through your life weighing up options and deciding what to do. Which football team will I support? What career will I pursue? Will I get married? Will I try for a baby? Will I drop out, wear a nose ring and promise myself never to own a lawnmower?

On the face of it, we seem to go through life in control of our own destinies. And yet, when you scratch the surface, it's not quite so simple. For instance, there may be a parasite in your brain controlling your behaviour. Or if conspiracy theorists are to be believed, you've been controlled since childhood to ensure you end up a productive tax-paying citizen. Certainly, you are being played by social media, but then you know that, don't you? The truth of the matter is your life is governed not so much by the stars but by random statistical fluctuations in a universe that is cold and unthinking. Or by Google. Cheering, isn't it? And look what happened with COVID-19. That particular bundle of random statistical fluctuations put paid to so many of your plans. But it's not too late – we will win the war against it and you will be able to take back control of your life and be truly free. Won't you?

Free will is an important concept in western civilisation. It's defined as the ability to choose between different options unimpeded. Many philosophers have tied themselves up in knots debating it. In some ways, it is the central question of philosophy.[1] Some contemporary philosophers gloomily complain about how little progress has been made on the issue over the centuries. Maybe philosophers have lost the (free) will to explore it further. We all have a strong sense of freedom, which makes us intuitively believe that we have free will. Spinoza felt that we are conscious of our actions (and interpret this as free will) but are unconscious of the causes by which our actions are determined. Also, if prior conditions determine everything that happens to us, then our future can be predicted with accuracy – we have no control over our future at all. Is it all predetermined? See what I mean by philosophers being troubled by all of this?

If you grew up in a stable family, with a parent who had a professional occupation, and if you are sent to a certain type of school and then on to a reputable university, all of these pre-conditions will most likely lead to you having a professional occupation and living a life predetermined by the circumstances you grew up in. If you are part of a religious community with strict laws, say, the Amish community in the US, then you will grow up to lead a life consistent with those laws. You will marry another Amish person. You will never own a car. On average, no free will. Some religions believe that we live in a world that is predetermined or predestined, where all events are determined in advance. Lutherans believe that Christians live a predetermined life, with salvation after you die being predestined for those who seek God. Calvinists take a more extreme view. They believe that God picked those who will be saved before the earth was even created. Hard luck if your soul wasn't picked – you might live a blameless life, but you still won't make it.

The German philosopher Friedrich Nietzsche, hero to many a disaffected teenager, didn't believe in free will at all and, judging from his writings, was annoyed by the range of teachings on the matter.[2] He put free will down to the pride of man (he didn't say much about women). He called the whole idea of free will 'crass stupidity'. Even though he declared 'God is dead', he wrote

FRIEDRICH NIETZSCHE (1844–1900), GERMAN PHILOSOPHER AND HERO TO TEENAGERS EVERYWHERE, WHO SAID, 'GOD IS DEAD' AND 'TO LIVE IS TO SUFFER'. HE ALSO SAID, 'WITHOUT MUSIC, LIFE WOULD BE A MISTAKE', REVEALING THAT HE HAD NEVER HEARD JUSTIN BIEBER.

about how free will absolves God (if there is one) of responsibility when humans misbehave: if humans are free, they can be deemed guilty by God if they sin. Nietzsche was also a fan of chance. He was of the view that much of what happens to us is governed by chance, as opposed to us deciding things. One of his big arguments was that 'if both humans and God actively will good things to happen, why is evil a constant in the affairs of man?' This question then led him to ask: where is this 'freedom of will' and why aren't we doing more with it?

And so, science enters the fray. Physicists are all about laws of nature – everything in nature can be predicted based on these laws. So all subsequent events, even those involving us humans, should be wholly predictable from first principles if we know the rules of the game: this was one of Newton's great insights. Science allows you to predict the future if you know the prior conditions. If you fire a cannonball of a particular weight, with a particular force, you can predict, using equations, exactly where it will land after you've fired it. The ability to predict what will happen (usually using mathematics) gave science primacy for many people. One problem is that in the spooky quantum world, predictions are made based on probabilities. This counters the notion of things being fully deterministic. Some philosophers have suggested that the quantum world and free will are somehow entangled, but trying to explain that is well above my pay grade. And some physicists believe in parallel universes, where every decision leads to two alternative universes.[3] So why worry? You're living multiple alternative lives somewhere else.

Neuroscientists have studied what it is that makes people make decisions and take action and agree that – guess what? – free will does not exist.[4]

A good illustration of this conclusion was an experiment performed in the 1980s by neuroscientist Benjamin Libet.[5] He asked people to choose a random moment to flick their wrist while measuring electrical activity in their brains. Specifically, he measured something called 'the readiness potential', which is brain activity in the lead-up to a voluntary muscle movement. It was well known that the readiness potential could predict the subsequent

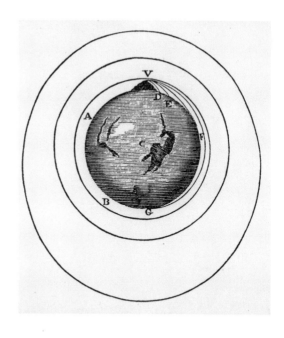

physical action (in this case, the wrist flick). Libet wondered if he could record this activity before the conscious intention to move. He asked the subjects to record the time when they felt they were about to move (as in, when they became conscious of the move that they were about to

'NEWTON'S CANNONBALL' WAS A THOUGHT EXPERIMENT THAT ISAAC NEWTON USED TO SHOW THAT IT MIGHT BE POSSIBLE TO FIRE A CANNONBALL SO FAR THAT IT WOULD NEVER LAND, INSTEAD GOING INTO ORBIT AROUND EARTH.

make). He noticed that brain activity preceded the conscious awareness that a movement would be made: the person's declaration of intention to move occurred after the brain had decided to move, which the person wasn't aware of. The person thought they were making a decision to flick their wrist (and to exert their free will), but the movement was being controlled subconsciously. How this subconscious control might extend into other behaviours is not known, and although both the design of the experiment and the interpretation are controversial, it remains a fascinating experiment.

Does the Libet experiment extend to social engagement? Say you're in a bar and see someone you like the look of and engage them in conversation: you

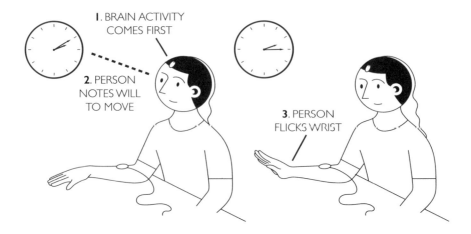

1. BRAIN ACTIVITY COMES FIRST

2. PERSON NOTES WILL TO MOVE

3. PERSON FLICKS WRIST

BENJAMIN LIBET'S EXPERIMENT ON FREE WILL. BRAIN ACTIVITY PRECEDES THE PERSON'S AWARENESS THAT THEY ARE ABOUT TO MOVE.

think you're making a decision to approach them for a chat and take the matter further, but perhaps your brain just fired a readiness potential. So it's not you that makes the decision at all; it's your brain. This is a hot topic in neuroscience, and the jury is still out because of issues with the design of some experiments.[6]

We spend much of our adult lives working in jobs we don't want to do and wasting our money buying things we don't need. We eat way too much even though we know that half the world is starving. If free will were real, surely we would make better decisions? How we make decisions is governed by a complex mix of external events and our own internal world. The mixture of influences on us is vast – chief among them are our evolved natures, our genetics, our hormones, how we were raised as children and the kinds of things we were exposed to and maybe even whether we have recently eaten. A Swedish study demonstrated that a hormone called ghrelin, produced by the digestive system when the stomach is empty, makes rats much more impulsive.[7] Studies have also shown that when we are hungry, we are inclined to make decisions that will lead to instant gratification, rather than weighing up the situation and making a decision based on a long-term gain.[8] You might think to yourself, 'I am making this decision by using my free will' whereas you are actually making it because you are hungry and ghrelin is making you change your behaviour.

Psychologists advise that before making a decision, you should keep a few things in mind. First, sleep on it: this will give added perspective. Second, never make a decision if you are feeling overwhelmed or low in energy: these feelings often give rise to a decision you will later regret. Third, it's probably best to make a decision on a full stomach.[9] And then there is the Irish proverb 'A man should never make a decision without consulting a woman.' Another version might be 'without consulting a scientist'. My wife Margaret is an outstanding biochemist. I often consult her on scientific issues (smart of me, huh?). Science tries to provide clarity for decision-making by using statistics, verifiable sources and conclusions based on evidence. Compare this to politicians, who might just write something on the side of a bus.

So even when we decide to act on something, the decision might be governed by how hungry or tired we are. But what if there were a parasite in your body controlling you? *Toxoplasma gondii* is a common parasite in cats,[10] and fascinating to microbiologists. It infects the brains of many species, including humans, all over the world, but cats are the only animals that can support the sexual stage of the parasite. To *Toxoplasma*, the cat is the love shack. Once a person or animal is infected (from the faeces of the cat, or from eating an infected animal), the parasite can remain in the host body for life in the form of latent cysts. These cysts are found in the brain, heart and muscle. When mice become infected, something curious happens: their behaviour changes dramatically. They become more reckless and are actually attracted to the smell of cats, which usually ends badly for the mouse. Darwin would love this example – a complex interplay between three species: cat, mouse and parasite. The mouse's behaviour has been modified by the parasite in order to promote transmission to the cat and also provide a tasty snack.

Humans infected with *Toxoplasma* are more prone to outbursts of aggression.[11] Those with a latent infection also do better in cognitive tests.[12] The level of infection also differs between countries: around 7 per cent of Irish

people are infected, whereas 67 per cent of Brazilians are.[13] Might this make Brazilians more hot-headed? Men and women also respond differently to the infection: men become less risk-averse but more dogmatic, while women become more outgoing.[14] Yet again, you might think what you're doing is based on your own free will, but in reality, you are the pawn of a parasite in your brain.

Or maybe a pawn in a world governed by statistical probability. When something happens to us, we often say 'What are the odds of that happening?' We are often amazed by coincidences or apparently random events determining our lives: 'If I hadn't picked up that newspaper on the DART and read about that job, I wouldn't have applied' or 'If I hadn't gone to that party, I wouldn't have met the person who I ended up marrying.' The road not taken might have made all the difference, but there is a probability of these things happening, and they may not be as unpredictable as you think.

Our amazement with coincidence largely comes from a poor understanding of probability. If you meet someone with the same birthday as you, you might say, 'Wow! That's a huge coincidence!' The probability of that happening is

1 in 365. And here's an interesting piece of maths: how many people need to be in a room for there to be a 50/50 chance that two will have the same birthday? The answer is 23 people,[15] which looks like a tiny number. But there are seven billion people on earth and, given that large a sample size, there is a large probability of outrageously unlikely things happening somewhere. If lots of people buy a lotto ticket, one will win – no surprise there. The surprise is only for the person who won.

What happened to Violet Jessup is extraordinary because she survived the sinking of three famous ships.[16] Violet was on board the *Olympic* when it crashed into *HMS Hawke* in 1911. The following year, 1912, she survived the sinking of the *Titanic*. Then in 1916 she survived the sinking of the *Britannic*. How could this be? Surely the probability of one woman being in three of the most famous sinkings in history is vanishingly small? It is, until you learn that she was a nurse who worked for the White Star Line and was assigned to these three ships.

What all this means is that things will happen to you based on probability, so just sit back and let them happen. You might change your fate if you actually do the lotto, or if you join a club in the hope of meeting new friends. Both appear to involve you expressing your free will, but they really aren't. You are drawn to the lotto because the adverts are aimed at you and you think you might have a chance and we could all do with extra cash. Choosing to do the lotto is almost out of your control. Your penchant for the lotto may be part of an addictive personality that you partly inherited from one of your parents. Addiction is, in fact, sometimes seen as evidence for free will, since it robs us of the capacity to make a choice, say, not to take a specific drug. However, another viewpoint is that people who are addicted are making choices every day to avoid the distress of withdrawal and to choose the pleasure of the drug. You might well make new friends at a new club, but that's highly likely because you'll be with like-minded people. Your interest in the club may have been preordained from your childhood. It was inevitable that you joined that club – your life history led you to it.

THE UNSINKABLE VIOLET JESSUP (1887–1971).
VIOLET SURVIVED THE SINKING OF THREE FAMOUS
SHIPS: THE *OLYMPIC*, *TITANIC* AND *BRITANNIC*. SHE
WAS A NURSE WHO WORKED FOR THE WHITE STAR
LINE, WHICH OWNED THE THREE SHIPS.

Apart from random events (which have a probability of happening) and decisions we make based on our history or perhaps our genetic makeup or maybe even a parasite we are carrying, how else are our lives controlled? Think about an average day. You get out of bed (after a solid seven hours sleep), choose what to eat (a fibre-based cereal with extra goji berries), choose what to wear (big meeting today, so a neat suit), go to the gym before work (because you've read it will make you more effective in your working day, not to mention that your Fitbit tells you to do so), go to work (to make money and achieve self-actualisation), have a glass of wine with friends on your way home (one glass only) and then binge-watch *Game of Thrones* to de-stress. You constantly make decisions as you go about your day, but on what are those decisions based? Many are based on advice you've read or heard. You can always choose not to heed the advice, of course, but most of us do, at least most of the time. We can live our lives by numbers if we choose:[17] we're told we need five portions of fruit and vegetables per day and if we don't get them, all kinds of awful things will befall us. We're told not to drink more than 14 alcohol units a week for women and 21 for men. We're told to get at least seven hours of sleep per night. We need 30 minutes of exercise, five times a week. These numbers are well supported by science in terms of being highly beneficial to us. It makes sense to follow them. Many millions of people try to follow them, usually falling off and then back onto the wagon. An alien looking at human behaviour when it comes to these health-based activities would conclude that the directive of a higher order is at play here, rather than people exercising free will.

If children are raised in a certain way, they will turn into adults with particular traits, governed as much by their upbringing as the choices they apparently make using free will. This belief, of course, lies at the heart of many religions. As Aristotle famously said, 'Give me a child until he is seven and I will show you the man.' (St Ignatius Loyola, who founded the Jesuits, plagiarised the phrase.) The resulting adult may well think that all their decisions have been conscious, when they were in fact preordained. In a study of 700 people from across the US aged between kindergarten age and 25, a significant correlation was found between the social skills of children and their success two decades later.[18] Children who could cooperate with their peers without prompting and who were helpful to others were far more likely to obtain a college degree by the age of 25 than those with limited social skills. The study shows that helping children develop emotional and social skills is key to future success. In another study, it was shown that daughters of working mothers went to school for longer and were more likely to have a job in a supervisory role and earn more money – up to 23 per cent more – compared to peers who had stay-at-home mothers.[19] Sons of working mothers pitched in more with household chores and also childcare later in life. And in a study that deserves the label *No Shit, Sherlock*, it was shown that the higher the parents' income, the better the child does in school, especially in a metric called the SAT score, which is the result of a standardised admissions test done by students who wish to go to university.[20] Another major predictor for gaining a college place was encouragement from parents at an early age. This is known as the Pygmalion effect – what one person expects of another can come to serve as a self-fulfilling prophecy.[21]

Advertisers exploit the fact that childhood influences can shape adult choices. A number of studies in the US show how children under the age of seven who are exposed to advertisements for fast food or sugary drinks develop a habit for these foodstuffs that becomes hard to break. If you give 3–5-year-olds identical foods, they will identify those in the McDonald's wrapper as tasting better.[22] The World Health Organization (WHO) has said that food giants are exploiting loopholes in regulation to advertise fast food to children through adverts on YouTube and Facebook.[23] A recent study in the UK found that 75 per cent of under-16s are being exposed to such adverts on social media.[24] It's a serious issue and the WHO has concluded that there is unequivocal evidence that exposure to fast food and sugary drinks in childhood is a major cause of the current obesity epidemic.[25] In the UK, products high in fat or sugar can only be advertised if the audience is at

least 75 per cent adult,[26] and a poll in Ireland recently concluded that 71 per cent are in favour of completely banning advertising fast food to children.[27] There is conclusive evidence of a causal link between fast-food marketing to children and childhood obesity. One former advertising executive who has since joined the effort to ban fast-food advertising said that 'junk food advertising has become a monster, manipulating young people's emotions and their choices'.[28] As yet, there are no rules governing the advertising of junk food in Ireland. Even worse, the consumption of fat and sugar can cause behavioural changes, including an increase in impulsivity,[29] which is likely to lead to yet more bad decision-making.

The battle to control advertisers has moved to social media – the directive of a higher order mentioned earlier. Social media is a relatively recent and exponentially growing influence on our lives. The overwhelming evidence is that we are ceding control of our lives to the machines we use to access social media sites. Our smartphones are insinuating their way into our lives and consciousness. They are having a greatly disruptive effect on our sleep. Again, we seem to have control over this (just turn it off), yet many of us are addicted to our iPhones and check them constantly, including when we should be asleep. An amazing 40 per cent of teenagers report checking their phones twice *during* the night.[30] This means severely disrupted sleep with all the obvious negative consequences, including increased risk of anxiety and depression. The most insidious way social media controls us is through advertising. The economic model of companies like Facebook and Google is blatantly obvious – and enormously lucrative. These companies collect data on their users and then sell it to advertisers who use it to target likely customers. It's the one thing advertisers have always wanted: to target their ads at the right person. This is a good thing, right? You get to see ads you actually want to see and buy products you really want. Well, not quite, as it turns out. There is mounting evidence that advertisers can determine all kinds of things about you – things you may not want them to know (such as, say, the fact that you like Italian food because you order a pizza online every Friday night).

You unwittingly reveal an awful lot about yourself on social media, including personality traits and political leanings. In a recent study, Facebook users' 'likes' were examined and, from that, people could be categorised into 'extrovert' or 'introvert'.[31] Ads were then tailored to each group. For instance, beauty companies sent ads to the introverts with phrases such as 'Beauty isn't always about being on show,' whereas to extroverts they sent phrases like 'Love the spotlight and feel the moment.' The campaign reached 3.5 million users, attracted 10,346 clicks and ultimately 390 purchases. Those who saw an ad tailored to their personality type were 1.54 times more likely to make a purchase. This technique is known as psychological mass persuasion. The selling of beauty products is one thing, but what if it's sending ads for gambling to people who might be at risk of gambling addiction? Or what if you're a Russian agent and you want to stir up trouble by sending ads or

MARK ZUCKERBERG, THE MAN WHO SINGLE-HANDEDLY DESTROYED ALL OUR LIVES BY INVENTING FACEBOOK, AT A CONGRESS HEARING IN OCTOBER 2019.

messages to people who you feel might be sympathetic? It's difficult to know what to do about all this. Once you've revealed who you are (which is what social media is all about), you are open to being exploited.

The issue becomes especially important when we look at democracy. There have been accusations that Russian intelligence officials manipulated people in the US via social media during the last presidential election, and that Cambridge Analytica also influenced the election. Cambridge Analytica was a British consulting firm that combined data analysis (largely gleaned from social media) with strategic communication during electoral processes. This company was paid £5m by the Trump campaign to help them target swing voters.[32] The Cambridge Analytica website boasted, 'We collect up to 5,000 data points on over 220 million Americans, and use more than 100 data variables to model target audience groups and predict the behaviour of like-minded people.' There is also evidence that Cambridge Analytica did work for the Leave campaign and UKIP ahead of the 2016 Brexit referendum. What is alleged to have happened is that Cambridge Analytica obtained data from millions of Facebook users, without their consent, and used that information to target them with pro-Brexit advertisements.[33] Facebook has since been fined $5 billion for its role in the data scandal, standing accused of not sufficiently protecting its users.[34] The controversy led to the closure of Cambridge Analytica in 2018. What they did is actually part of a growing trend for political groups to use digital political campaigning – targeting people with highly sophisticated messages – as a key strategy. And Cambridge Analytica is not the only culprit: both Barack Obama and Hillary Clinton employed behavioural profiling companies. But the big question is: does it work? Simon Moores, a world expert on cybersecurity, is of the view that behavioural modelling that involves big data analysis has passed an inflexion point. He says we can 'look forward to a future that's made up of equal parts Orwell, Kafka and Huxley'.[35] When you go to the ballot box, then, will you be voting with no free will at all?

When we read about the likes of Cambridge Analytica, we have to wonder about the level of control we have over our lives. It seems that we're being manipulated by algorithms. iPhone users of the world, unite – you've nothing to lose but your devices. Alternatively, you could try net-based services that avoid the metadata miners and free yourself from their clutches. Whatever about the decisions we make and the lives we lead, many things in our lives are for definite beyond our control. We blame fate for events such as serious illness, loss of loved ones, accidents, macro-economic downturns, famine and war. We can, of course, try to limit the risk of these things happening if we, say, look after our health and follow advice from experts, but we need to escape the hold social media has over us. We need to use science to help us make the right decisions, including when it comes to the threat posted by COVID-19. **The bottom line: don't get married to a stranger in Las Vegas while jet-lagged after bingeing on social media. In all probability we can regain control over our lives and a brighter universe will be around the corner. Now, keep reading. Go on, go on, go on, you know you want to.**

CHAPTER 2

WHY WON'T YOU GET VACCINATED?

—

'I knew they wouldn't play it when I wrote it.'

—

Ian Dury, who had polio, on his song 'Spasticus Autisticus'

HAVE TWO SONS, Stevie and Sam, and they have both been vaccinated with every vaccine available. It's simple. I love them and want to protect them. No doubts, no fears.

If you really want to annoy an immunologist, tell them that you haven't vaccinated your child. Vaccines against infectious diseases have saved more lives than any other single intervention in medical history.[1] They prevent an amazing two to three million deaths per year around the world. This is a scientific fact.[2] So is the fact that before vaccination, around 500,000 people caught measles in the US, with three in ten having permanent hearing damage as a result.[3] And yet growing numbers of parents and guardians are refusing to vaccinate their children. The situation has become so bad that the WHO, which has the health of people as its main concern, listed vaccine hesitancy as one of the top ten threats to global health in 2019, as dangerous to our health as pandemic influenza, Ebola virus and antibiotic resistance.[4] We can add the latest member of the virus rogues' gallery, SARS-CoV-2, the coronavirus that causes COVID-19, for which we desperately need a vaccine.

How could vaccine hesitancy have happened in the first place? How could one of the greatest advances in medicine have become so problematic for many people, especially when the evidence in favour of vaccination is so overwhelming? How can we convince parents who are reluctant to vaccinate their child that they risk not only their own children becoming sick, but also that they are putting others at risk? And what difference has COVID-19 made? Will the greatest pandemic since the Spanish flu of 1918 win over hearts and minds, and decrease vaccine hesitancy among worried parents?

In some ways, a distrust of vaccines is understandable. A young mother goes into her GP's surgery, carrying her lovely, healthy baby. She tells her GP, whom she likes and trusts, that she doesn't want a needle stuck into her baby, who isn't sick. She's heard scary stories. She wants none of it. At the height of the MMR (measles, mumps and rubella) vaccine scare, friends of mine asked if they should vaccinate their children. Friends with law degrees and business degrees. My answer: unequivocally, yes. Where's the evidence that made me say yes, I hear you ask? Well, here it is.

Take measles. This disease is caused by a highly contagious virus. Initial symptoms are a fever (which can run as high as 40°C and cause convulsions in children), a runny nose, a cough and inflamed eyes. A flat red rash then appears and spreads all over the body. Common complications include diarrhoea and ear infections. Less common ones include blindness and death; One to two in 1,000 will die.[5] Nine out of ten people sharing a living space or a school with an infected person will catch measles. In 1980, 2.6 million people died of measles. They were mostly under the age of five. By 2014, following global vaccination programmes, this number fell to 73,000.[6] The introduction of the vaccine has had a remarkable effect. In the US, before vaccination, the annual rate of measles stood at 3–4 million cases per year. After vaccination, this number fell to almost zero.[7] All that sickness, lifelong complications and even death, prevented by a simple jab in the arm.

Or take polio, another viral disease. Infected people can have minor symptoms that clear up quickly, including sore throat and fever. However, for one person in every 150 people infected, the virus enters the nervous system, where it can wreak havoc. Initial symptoms include headache, back pain, lethargy and irritability. Some people will go on to develop paralysis, with muscles first becoming weak and floppy before complete paralysis. The

WELLBEE WAS USED TO PROMOTE PUBLIC
HEALTH IN THE US IN THE 1960S. HERE
HE IS ENCOURAGING THE UPTAKE OF
THE POLIO VACCINE. BRING WELLBEE BACK
TO HELP FIGHT COVID-19!

virus is usually spread in faecal matter or via mouth-to-mouth transmission. The famous rocker Ian Dury contracted polio at the age of seven from, he believed, a swimming pool in Southend-on-Sea during the 1949 polio epidemic. In endemic areas (i.e. areas where the virus is common) it infects everyone in the population. It was a disease that every parent feared. Author Richard Rhodes has written, 'Polio was a plague. One day you had a headache, and an hour later you were paralysed. Parents waited every summer to see if it would strike. One case turned up and then another. We all stayed indoors, shunning other children. Summer seemed like winter then.' Again, the introduction of the polio vaccine had a remarkable effect. Before vaccination in the US, there were around 15,000–20,000 cases of paralytic polio per year. After vaccination? That number fell to fewer than ten.[8] No more people were becoming paralysed. No more summers like winters. In 2002, Europe was declared polio-free, and that remains the case. Today, only three countries are not free of polio: Pakistan, Afghanistan and Nigeria.

Clearly, then, if safe and effective vaccines are developed, they can spell the end for infectious diseases that were a cause of much fear, suffering and death. So what are these wondrous things called vaccines? A vaccine is defined as a biological preparation that provides immunity to disease. The term 'immunity' comes from the Latin word *immunis*, meaning 'exemption'. In Roman times this usually meant an exemption from paying taxes, which was granted to certain Roman citizens (for example, returning soldiers). In the case of infectious diseases, immunity means exemption from getting the disease again. This had been noticed in ancient times. After someone got sick from a disease, they rarely got it again, and so would care for those who had the disease for the first time. The first written description of the concept of immunity might have been by the Greek historian Thucydides, who in 430 BC wrote that when a plague struck 'the sick and dying were tended by the pitying care of those who had recovered, because they knew the course of the disease and were themselves free from apprehensions. For no one was ever attacked a second time.' This was believed to be magical or God-given. An

THE REMARKABLE LADY
MARY WORTLEY MONTAGU
(1689–1762). SHE ADVOCATED
SMALLPOX INOCULATION
USING PUS FROM SMALLPOX
LESIONS. THIS WAS AN
IMPORTANT FORERUNNER
TO THE WORK OF EDWARD
JENNER.

early clinical description can be found in the writings of Islamic physician al-Razi, who described smallpox and how exposure to smallpox conferred lasting immunity. Smallpox was feared because it was highly contagious, killed one third of people who contracted it, and badly disfigured another third (and left the final third unharmed because their immune systems fought it off effectively).

The effort to prevent smallpox is important for the history of vaccines and the science of immunology, because it revealed a way to prevent infections. Around 1000 CE the Chinese began using dried crusts from skin lesions of patients with smallpox, which, when inhaled, provided some protection. Inoculation (meaning using a needle to insert material from smallpox lesions into the skin) was used in India and East Africa and introduced to the West in 1721 by Lady Mary Wortley Montagu.[9] Lady Mary was a remarkable woman. She was the daughter of a duke and had been due to marry an Irish aristocrat named Clotworthy Skeffington. Poor Clotworthy was dumped, and Mary eloped with Edward Wortley Montagu, who became the British ambassador to Turkey. While in Turkey, she recorded how smallpox inoculation was practised; the procedure involved taking pus from a smallpox blister in a mild case of the disease and then applying it to a scratch on the skin of an uninfected person. She had two of her own children inoculated in this way. To bring inoculation to the attention of people in England, she gave seven prisoners who were awaiting execution in Newgate prison the chance to undergo inoculation instead of execution. All seven survived and were released. The disease would have been close to her heart as she had a brother who had died of smallpox and she herself had survived it. Yet in some cases this method of inoculation actually *gave* people smallpox, because it sometimes contained live infectious virus.

In 1798 Edward Jenner, a Gloucestershire doctor, tried a far safer method by deliberately infecting people with cowpox, which caused a mild infection but showed remarkable protection against smallpox. The idea of using cowpox in this way probably originated in the observation that milkmaids (as women who milked cows were called at the time) often had beautifully smooth skin. This was put down to the fact that they rarely got smallpox and so didn't have the so-called 'pockmarks' on their skin, but yet would have caught cowpox from the cows they were milking. Might having cowpox somehow have protected the milkmaids from contracting smallpox? In fact, people who had cowpox usually cared for people with smallpox, since they were known to be unlikely to catch the disease. This approach of using cowpox to protect against smallpox was tried by at least five other investigators, including a farmer named Benjamin Jesty, who was a neighbour of Jenner's and who might have given him the idea.[10] When Jenner subsequently became famous for vaccination and had been granted £30,000 as a reward, Jesty sought compensation and was finally given two golden lancets as his reward. The credit had gone to Jenner partly because of an experiment he performed on an eight-year-old boy called James Phipps. Jenner scraped pus from the cowpox blisters on the hands of a milkmaid called Sarah Nelmes, who, history records, had caught cowpox from a cow called Blossom. He inoculated Phipps with the pus, which gave Phipps a mild fever. Jenner then injected Phipps with material from smallpox lesions (which would have been used for inoculation against smallpox) and the boy developed no symptoms of any kind. Normally that would have caused mild symptoms of infection.

EDWARD JENNER (1749– 1823) VACCINATING A BABY AGAINST SMALLPOX. JENNER GETS THE CREDIT FOR AN EXPERIMENT THAT INDICATED THAT COWPOX MIGHT PROTECT AGAINST SMALLPOX.

Jenner's key contribution here was the demonstration that cowpox pus could be used from one human to another and that the boy was protected from disease when challenged. Jenner used the term 'vaccination', from the Latin *vacca*, meaning 'cow'. Rather unexpectedly, scientists now think that Jenner might have used horsepox that had infected a cow – something Jenner himself believed. So perhaps we should call it 'equination'? He followed up the Phipps study with 23 more cases, including his 11-month-old son Robert. This is an important part of medical science: to repeat an anecdotal observation with a study in multiple patients. Whatever the original source of the cowpox vaccine, vaccination was adopted widely in England. Jenner became famous all over Europe. The Empress of Russia sent him a diamond ring out of gratitude. Napoleon said he 'could refuse this man nothing' even though France and England were at war.

In a forerunner of the modern anti-vaccination movement, many spoke out against the smallpox vaccination.[11] Clergymen felt that smallpox was a God-given fact of life and death, and any attempt to subvert this divine intention was blasphemy. Some religious people felt that smallpox was a force sent by God to cull the poor. Doctors also joined this nascent anti-vax movement. Many made a good living from quack treatments for smallpox and vaccines threatened their livelihoods. Strange reports began to appear in medical journals, telling how vaccination could transmit bovine traits, such as children making mooing noises and running around on all fours. The ridiculous contention that vaccination turned children into cows became widespread.

In 1906 a woman called Lora Little, who was described as a 'natural therapist', claimed that vaccination was a scam set up by doctors, vaccine-makers and the government. She described 300 cases where vaccination against smallpox had been harmful, including the tragic case of her seven-year-old son who died after being forcibly vaccinated. (He actually died of diphtheria.) In England, notable people spoke out against vaccination. Even George Bernard Shaw was against it, describing vaccination as 'a peculiarly filthy piece of witchcraft'. In another precursor to what is happening today, parents who refused to vaccinate against smallpox were fined or sent to prison. So anti-vaxxers are not new, and neither are attempts to deal with them.

Following Jenner's success, many other vaccines were developed. In the 1880s French scientist Louis Pasteur introduced vaccines for chicken cholera and anthrax, infectious diseases that afflicted farm animals. It was Pasteur who suggested in 1891, in honour of Jenner, that the term 'vaccination' be more widely used to describe inoculation against infectious diseases. Other vaccines soon followed: in 1884 for rabies, in 1890 for tetanus, in 1896 for typhoid fever and in 1897 for bubonic plague, also known as the Black Death. That disease had been the scourge of Europe, killing as many as 60 per cent of the population in the fourteenth century, but a vaccine finally vanquished it. From the late 1800s, vaccination became a matter of national

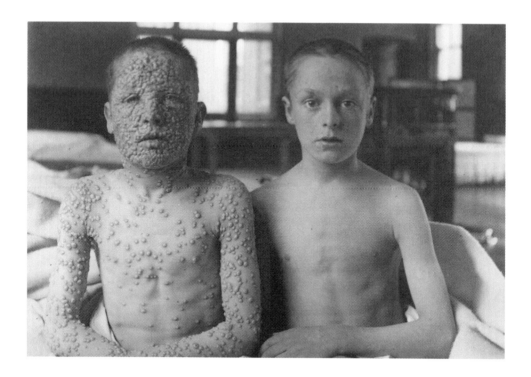

WHAT HAPPENS WHEN
YOU DON'T VACCINATE
AGAINST SMALLPOX. THIS
PHOTOGRAPH WAS TAKEN
IN THE EARLY 1900S BY
DR ALLAN WARNER. THE
BOY ON THE RIGHT WAS
PROTECTED AGAINST
SMALLPOX BECAUSE HE WAS
VACCINATED IN INFANCY.

pride, with countries boasting about protecting their people from horrible diseases. In the twentieth century, new vaccines came thick and fast: for tuberculosis, or TB (another scourge of many counties, including Ireland where it killed at least 10,000 people per year in the early part of the twentieth century), for diphtheria, scarlet fever, yellow fever, influenza, polio, measles, mumps, rubella, meningitis and hepatitis B. One by one, diseases that killed millions were beaten by the power of vaccines.

It's abundantly clear why vaccines are seen as the greatest contribution to medicine. Following on from Jenner's work, the scientific question became: how do these wondrous things work? Inoculation with smallpox was viewed as a folk remedy of uncertain provenance. The study of what Jenner and Pasteur had achieved gave rise to the field of immunology. The efficacy of cowpox as a preventive against smallpox provided the first clue as to how a vaccine might work. We now know that the cowpox virus is similar to the smallpox virus and yet doesn't cause that disease. The similarity means that when injected with cowpox, the body mounts an immune response to it and clears the mild infection. When someone is subsequently infected with smallpox, the immune system has been trained by its prior exposure to cowpox and kills the smallpox, because it recognises the parts of smallpox that are similar to cowpox. If the body hasn't seen cowpox, smallpox will run amok as the immune system hasn't been trained to recognise and kill it, causing disease. It's a bit like a nightclub where, let's say, boisterous fans of a football team wearing their team colours try to gain entry. The bouncers stop them. More fans then turn up wearing the same colours as the previous fans, but this time perhaps carrying weapons. They are immediately stopped from gaining access because the bouncers recognise them by their team colours. Cowpox and smallpox wear the same colours (meaning they have similar components) but smallpox is better armed and can cause more severe disease. Cowpox and smallpox are in the same family of viruses and so can be recognised by the same parts of the immune system.

Vaccines today are mainly of two types: dead or inactivated infectious organisms (weakened football fans wearing their team's shirt), or purified products from them (the shirt alone). Pasteur was the scientist who came up with inactivation as a method. He was studying chicken cholera, an infectious disease in the poultry industry at the time. In one experiment, he infected chickens with a batch of cholera, mixed into a broth, that had been left out and had spoiled. When he then tried to infect the chickens with fresh cholera, he noticed that they were protected. The cholera bacteria that had gone off had been weakened in some way and no longer caused disease but had components in common with the regular bacteria: it trained the immune system to respond to the more virulent bacteria. The first vaccines were therefore similar to the chicken cholera that had spoiled. They were somehow inactivated, or as it is usually termed 'attenuated', by chemicals or heat. These include vaccines against polio, hepatitis A, rabies, yellow fever, measles, mumps, rubella, influenza and typhoid. Jonas Salk inactivated the polio virus with the chemical formalin, whereas Albert Sabin discovered in infected animals a weakened version of the polio virus, which was less toxic. Both of these worked well against polio. The vaccine against typhoid was invented by Almroth Wright, who had studied medicine in Trinity College Dublin. It saved tens of thousands in World War I: before the vaccine was introduced, more soldiers had died of typhoid than in combat. For TB, a vaccine termed BCG (named after its inventors Calmette and Guerin; the B stands for 'Bacillus', the genus of bacteria that TB belongs to) has been used for decades, and it was introduced into Ireland by Dorothy Stopford Price in the 1950s, again saving many lives. BCG also causes a non-specific boosting of the immune system, meaning that it can put up a barrier (involving immune cells called monocytes) that can repel other infections, including viruses (e.g. the measles virus) and possibly SARS-CoV-2. BCG may well be useful as a way to protect against COVID-19, and at the time of writing, is undergoing testing.

A wide range of vaccines use component parts from the infectious agent. This can include inactivated toxic components (called toxoids) and includes vaccines against tetanus and cholera. Subunit vaccines comprise proteins from the infectious agent and include vaccines for hepatitis B, influenza and human papilloma virus (HPV) – the latter protects against cervical cancer, which is caused by HPV.[12] Influenza is a major focus for health authorities as it can cause death in the vulnerable, mainly the old, sick and very young. There are four types of influenza – A, B, C and D. All have proteins in their coats called hemagglutinin (H) and neuraminidase (N). There are different types of H and N and the virus can change from season to season, hence each flu season might have a different vaccine. Much effort is going into the search for vaccines for diseases like malaria (caused by the parasite *Plasmodium falciparum*) and AIDS (caused by the human immunodeficiency virus). These are proving difficult to vaccinate against, although there is some progress, particularly with malaria.[13] One of the most recent vaccine successes has been against Ebola virus, which

LOUIS PASTEUR (1822–1895), FRENCH MICROBIOLOGIST AND CHEMIST WHO INVENTED VACCINES AGAINST RABIES, ANTHRAX AND CHOLERA.

causes a highly lethal disease in parts of West Africa, with an overall mortality rate of up to 90 per cent. A major effort to develop a vaccine began following outbreaks in 2013 due to the high mortality rate and symptoms, which include internal and external bleeding and organ failure. A vaccine against Ebola was developed in 2015, and new ones are still in development.[14] The COVID-19 pandemic led to an unprecedented effort to find a vaccine, with at least 41 candidates in development in April 2020.[15] Every possible strategy is being tested: dead virus, live weakened virus and also components from the virus. The components include the protein in the spikes of the virus, which it uses to penetrate lung cells. Even RNA from the virus that encodes the spike protein is being tested. If it is injected into your arm muscle, your body then makes the spike protein and your immune system can then make antibodies that will bind to the spike protein like Blu-Tack, bunging it up and stopping the virus entering cells. These antibodies will hopefully protect you when the real virus comes along. It takes time and it is unlikely that a vaccine will be ready before 2021. This is because vaccines have to be carefully checked to see if they have any side effects, and then checked again in trials to see if they actually work.

Although the main part of a vaccine is a microorganism that has been weakened or a part thereof, another important part is called the adjuvant. This is a chemical that can boost the immune response, and most vaccines don't work without it. Adjuvants are akin to jump leads to get a car engine going. Widely used adjuvants include the chemical aluminium hydroxide (Alum).[16] Other examples include a chemical called monophosphoryl lipid A (MPL), which is used in the hepatitis A vaccine. These agents will stimulate the immune response, and in the case of MPL this involves the triggering of an immune system protein called TLR4, which boosts immune cell activation. Research into new adjuvants aims to boost vaccines for diseases yet to be conquered including AIDS and malaria, and there is great progress in determining what parts of the immune system might need boosting to enhance vaccine efficacy. New adjuvants are also being tested with vaccines for COVID-19, including an adjuvant called AS03, which might hold particular promise.

The use of adjuvants or other additives has added to concerns that vaccines can be harmful. The decrease in vaccine use is known as vaccine hesitancy, which is defined as a 'delay in acceptance or refusal of vaccines despite the availability of vaccination services'. It has been reported in more than 90 per cent of countries in the world. In the UK, for instance, coverage of the MMR vaccine has dropped to 91.2 per cent and is at its lowest level since 2011–12.[17] There is no doubt that vaccines can be harmful in some individuals although the incidence is rare; and given the overwhelming benefits to humanity of vaccination, adverse events are not sufficient to prevent vaccine use. Obviously, though, any harm caused by a vaccine can be devastating for the affected person or parents of a child who had a bad reaction. So what is the exact situation regarding harmful vaccines?

A recent study revealed how rare injury is. Over the past 12 years, 126 million doses of the measles vaccine have been given in the US. During that period, 284 people filed claims of harm from those immunisations, through a proactive federal programme created to compensate people injured by vaccine. Of those claims, 143 were compensated, giving a 1 in 818,119 chance that the measles vaccine will harm you. The data come from the National Vaccine Injury Compensation Programme, a no-fault system that began in 1988.[18] This contrasts with a 1 in 500 chance of an unvaccinated child dying of measles, rising to 1 in 10 for a malnourished child. It also contrasts with a 3 in 10 chance of ear infections leading to permanent hearing loss.[19] Overall, when billions of vaccine doses have been given to hundreds of millions of Americans, only 6,600 people needed to be compensated for harm, with $4.15 billion dollars being paid out in compensation. Meanwhile, the Center for Disease Control in the US has estimated that vaccines have prevented more than 21 million hospitalisations in the US, and 732,000 childhood deaths.[20] All those children are still alive because of vaccines.

Vaccines can still have minor side effects – a sore arm, or low-grade fever. Other slightly more serious side effects include seizures in response to the MMR vaccine, which occurs in 1 in 4,000 children,[21] and with the HPV vaccine against human papilloma virus, which in one study of almost 200,000 girls caused 24 cases of fainting.[22] One reason why there has been a concern is that sometimes children become ill soon after vaccination, but in the vast majority of cases this is a coincidence: one example of this coincidence is Sudden Infant Death Syndrome (SIDS), which cannot be linked to vaccination – it happens, at the same rate, in unvaccinated children.

The anti-vaccination movement got a major booster shot in the arm in 1998 when Dr Andrew Wakefield published a paper, which stated that the MMR vaccine was linked to autism.[23] This study is likely to have been responsible for serious illness and even death in many children, because it scared parents into not vaccinating their children. Numerous flaws were found in the study including the small numbers of patients studied. Wakefield took money from solicitors who wanted to sue vaccine manufacturers, constituting a major conflict of interest for his publication linking the MMR vaccine to autism. *The Lancet* retracted the study in 2010 and the General Medical Council (GMC) in the UK struck Wakefield off. The council concluded that Wakefield had acted against 'his patients' best interest' and been dishonest in his research. Elements of his publication had been falsified and the *British Medical Journal* stated that his work was 'an elaborate fraud'. Wakefield was also found to have used children who showed signs of autism as guinea pigs, putting them through invasive procedures including colonoscopies and painful lumbar punctures. He went to a children's party and paid some children £5 for blood samples. Many scientists further explored the possible link but study after study was then published showing no link between the MMR vaccine and autism. The Center for Disease Control, the American Academy of Pediatrics (which has 62,000 paediatricians as members) and the FDA (the Food and Drug Administration – the US government agency that must give ultimate approval to a new drug) has stated that the MMR vaccine is safe.[24]

Fears against vaccines were also stoked by Robert F. Kennedy, son of Bobby Kennedy, who stated that people were becoming sick from a vaccine additive, thimerosal, which contains mercury. Kennedy has an autistic son. He and actor Robert De Niro offered a prize of $100,000 for anyone publishing a study proving thimerosal to be safe. This is almost impossible to achieve – could you prove that water is safe? One study in 2014 reviewed ten separate studies involving over 1.25 million children and concluded that there was no relationship between the MMR vaccine or thimerosal and developing autism.[25] Good enough for me. As for Wakefield, he continues to defend his findings and has reportedly given lectures on the *Conspire-Sea* cruise ship, sharing the platform with crop-circle obsessives, a woman who claims to have visited Mars and a man who insists that he has died and been reborn three times.[26]

The anti-vax movement continues to make its case against vaccines. This has led to a re-emergence of measles, including cases in Ireland, which more than doubled last year.[27] UNICEF has reported that 169 million children globally missed out on their first dose of the measles vaccine between 2010 and 2017. Cases of measles have increased in 98 countries. Why would parents hesitate when it comes to vaccines? Internet-based advocacy against vaccination is likely to sow fear and doubt. Should vaccinating your child be a legal requirement, as it was in the early twentieth century for smallpox? Some argue this infringes human rights, but others state we do this kind of thing already in different contexts. Seatbelts are compulsory, even though they might harm you by rupturing your spleen. It is public health for the public good, just like the drink-driving laws.

ROBERT F. KENNEDY (SON OF BOBBY KENNEDY) AND ROBERT DE NIRO. WHAT IS IT ABOUT ROBERTS THAT MAKES THEM VACCINE-DENIERS? YOU TALKIN' TO ME, ROBERT?

Let's look at 'herd immunity' to explain why parents should vaccinate their children. Over 95 per cent of people need to be vaccinated to stop the measles virus – it will then have nowhere to hide.[28] If you don't vaccinate your child, you are putting other people at risk – people who have a weakened immune system because they are on immunosuppressant drugs (given to people who have had organ transplants or are being treated for an inflammatory disease like rheumatoid arthritis) or because they have diabetes or heart disease or simply because they are an older person, since like everything else, their immune system gets less agile as they age. These vulnerable groups can of course be vaccinated, as happens with the flu vaccine, but to decrease the risk of them becoming infected, herd immunity provides another safeguard. COVID-19 is especially severe in older people and people with these underlying conditions,[29] making vaccination and herd immunity all the more important to protect the vulnerable against SARS-CoV-2.

What can be done to educate parents? The best way is for a doctor to approach a reluctant parent or guardian with empathy and humility (as should be the case for every doctor–patient interaction). The doctor might begin with: 'We all love our children. We all want the best for them. I understand that you're not trying to hurt your child. Now let's talk.' Tone of voice is everything. If you ask 'Why do you think that?' make sure you don't say 'Why do you think *that*?' Concerns should be acknowledged. Personal stories should be shared as these can create an emotional resonance. Roald Dahl, the beloved children's writer, described how his seven-year-old daughter caught measles. He wrote about how she seemed to be recovering, sitting up in bed. He began teaching her how to make farm animals from pipe cleaners, and then he noticed how she had trouble coordinating her finger movements. One hour later, she was unconscious, and 12 hours later she was dead. This happened a year before the measles vaccine was developed. Telling parents this story or showing them the essay Dahl wrote after her death, is likely to work more effectively than facts on those more likely to respond to fear than to reason.

But others disagree: fear might make things worse. One study in 2015

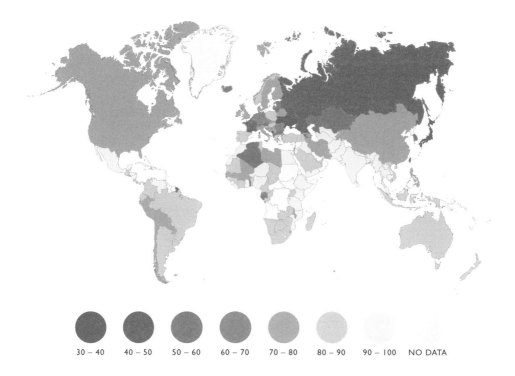

| 30 – 40 | 40 – 50 | 50 – 60 | 60 – 70 | 70 – 80 | 80 – 90 | 90 – 100 | NO DATA |

split 315 people into three groups.[30] They showed one group information debunking the link between MMR vaccines and autism. They gave another group scientific information not related to vaccines, and they showed the third group pictures of children suffering from mumps, measles or rubella. When asked, the third group viewed vaccines more favourably than the other two groups, which was seen as a good result. But in another study, people were shown scary pictures of infections and tragic stories, which put them off vaccines. There is some evidence that an action that might cause harm (say, giving a child a vaccine) is viewed as worse than inaction causing harm, as the latter is seen as fate and not the fault of the parent.

SHARE OF PEOPLE WHO BELIEVE VACCINES ARE SAFE (%). SOURCE: WELLCOME GLOBAL MONITOR.

The aim then is to try to overcome vaccine hesitancy using multiple approaches. It's vital for the doctor or scientist never to be judgemental. One helpful tactic is to state consensus among scientists, for example,

'Ninety per cent of medical scientists agree that vaccines are safe and that all parents should vaccinate their children'[31] or 'Children are 10,000 times more likely to be brain-damaged by measles than by vaccination.' These kinds of statements avoid repeating myths in an attempt to debunk them.

Anthony Fauci, the director of the National Institute for Allergies and Infectious Diseases (NIAID), who is on the front line in the fight against COVID-19 in the US, recently told a US Congress hearing that the main issue causing vaccine hesitancy is misinformation. Vaccine-hesitant parents are more active in searching for information online than vaccine-compliant parents. This has led Facebook to announce that groups and pages sharing anti-vaccine misinformation will be removed from its recommendation algorithm. Medical professionals who are trying to increase vaccine compliance see this as a crucial move since it will promote the spreading of evidence-based information explaining the benefits of vaccination. And what does that evidence-based information tell us? Overall, serious allergic reactions occur in one in 1 million doses of vaccines. Some of the specific concerns raised can be addressed as follows:[32]

Getting too many vaccines will overwhelm my baby's immune system. The real situation: it's true that children are being given more vaccines than before and sometimes in combination, but overall the amount of material in each vaccine is a lot less than before so in fact the overall exposure is less. The amount of material in vaccines is in fact trivial when compared to what children get exposed to in the natural world every day.

My baby's immune system is immature so it's safer to delay some vaccines. The real situation: this is not true – delaying vaccines increases the risk of infection and, in the case of the MMR vaccine, might increase the risk of febrile seizures.

What's in vaccines anyway? Don't they contain chemicals and toxins? The real situation: vaccines do contain things like aluminium and formaldehyde but at levels much lower than what your child will pick up in the environment.

The side effects of some vaccines seem worse than the actual disease. The real situation: vaccines have all been through rigorous safety testing, which can take 10 to 15 years. The FDA in the USA and the European Medicines Agency (EMA) in Europe monitor and scrutinise all the tests: only then, if all is well, will they grant approval. No company would invest in a product that would cause more serious health problems than it prevents.

If all else fails, we might need to make it illegal not to vaccinate. In the US, parents have to get an exemption, on medical or religious grounds, if they want to send their unvaccinated child to school. This has been shown to definitely increase the rate of vaccination. France has made 11 vaccines mandatory in order for children to be enrolled in school. Although draconian, perhaps that is the best way to protect our children. COVID-19 will make this all the more likely, given the devastation it has wreaked, both for human health and the world economy. The consequences of a serious infection for which there is no vaccine are now obvious – mass isolation, disruption of normal activities and an enforced but accepted cessation of human rights for the common good. Because of their great success rate at preventing millions of people from becoming sick and dying, work into vaccines continues apace. There are massive efforts afoot to develop new vaccines for all kind of diseases, including cancer. Imagine that – a vaccine to prevent cancer, which is already there for cervical cancer with the HPV vaccine and could be a reality for many other cancers. The Center for Disease Control in the US says that the HPV vaccine could prevent 90 per cent of the 32,100 cases of cervical cancer. All those people cancer-free. Many infectious diseases remain high on the agenda, as mentioned above, with a big effort into developing vaccines against malaria and AIDS, and new and better vaccines against diseases like TB and, of course, COVID-19. Progress is happening and the future looks bright. **Bottom line: my children were fully vaccinated. All health agencies in all countries agree: vaccinate your child.**

CHAPTER 3

WHY ARE NEW MEDICINES SO EXPENSIVE AND WHO SHOULD BEAR THE COST?

—

'I'm so mean, I make medicine sick.'

—

Muhammad Ali

SPEND A LOT of my working life doing medical research. I work on inflammation in the body, a process that protects us from harm and fixes our bodies when they are damaged. The problem is, inflammation can sometimes turn on us and give us painful and debilitating diseases such as rheumatoid arthritis, Crohn's disease, multiple sclerosis and Parkinson's disease, to name a few. All these diseases involve our bodies becoming inflamed and badly damaged, with symptoms of that damage being all too real. The good news is that progress is being made in the effort to find medicines to treat these diseases. Beyond these diseases, look what is happening with COVID-19, where a huge number of new vaccine candidates, anti-viral and anti-inflammatory therapeutics were rapidly developed and tested. This illustrates how, when galvanised, scientists and drug developers can put a huge amount of effort into finding a solution. It's a slow business, but there is hope.

On the face of it, the future of medicine looks promising for all of us. Hundreds of thousands of scientists and doctors have been working for decades with one goal in mind: to find new treatments or to come up with ways to prevent diseases that afflict humanity. Right now, new drugs are being approved for a range of diseases for which there were either ineffective treatments or no treatments at all. With many more in the pipeline (a term that both oil companies and pharmaceutical companies use), all kinds of treatments, ranging from new tablets to injections to modification of troubled genes to replacement of organs, are now within reach. But one question won't go away. It's the drugged elephant in the room, if you will. Who will pay for all these advances? Will you? Can you? Should the government bear the brunt of the cost? What will it mean for our health services? How can we put a price on your life or the life of your child?

Everyone's life follows a different course, full of ups and downs and challenges of many kinds. They are simply life's vicissitudes. But one of the most unfair concerns our health: some people get sick with a serious disease and some don't, although all of us eventually suffer from the ultimate vicissitude – ageing. In ancient times, before we had any scientific understanding of diseases, people struggled to rationalise the seemingly random nature of who would be struck down and who wouldn't. Did someone put a curse on you? Did the gods do it because you were a bad person? All kinds of superstitions prevailed in an attempt to ward off illness. Old Irish superstitions included tying a bunch of mint around your wrist to stave off infections. If someone died of a fever, a flock of sheep should be run through the house to protect the other inhabitants. A sick person's bed should be placed north to south, or perhaps, as Shane McGowan would have it, if you're Cuchulainn, put a glass of punch at your feet and an angel at your head. St Blaise was the patron saint of sore throats (and of wool combers, who must have suffered from sore throats). Our ancestors were desperate, invoking a multitude of patron saints of diseases: St Agatha for breast cancer; St Alphonsa for diseases of the feet; St Christina the Astonishing (she must have been great) for mental illness; St Ursicinus for a stiff neck. And although somewhat oblique, St Olaf was the patron saint of difficult

SAINT BLAISE, PATRON SAINT OF SORE THROATS AND WOOL-COMBERS (WHO MUST HAVE HAD A LOT OF SORE THROATS).

marriages (which present a number of health risks) and St Expeditus was the patron saint of procrastination (whom I prayed to regularly while I wrote this book). One has to wonder if these saints were the first medical specialists. Some religious practices might also originate in the fight against illness, including fasting, which can bring many health benefits,[1] and not eating pork, which might protect against parasitic worms.

Of all the things that can happen throughout our lifetime, we fear illness the most. It brings enormous worry, and it is different from other concerns, like getting a job or finding a partner, which we might be able to control. A diagnosis of any type of cancer or neurodegenerative disease (of which there are many, including Parkinson's and Alzheimer's) or a host of inflammatory diseases can provoke depression. Depression will affect 25 per cent of cancer patients, 50 per cent of people with Parkinson's disease and one in three heart-attack victims.[2] This can happen because of the underlying disease process affecting our brains (as is the case with inflammatory conditions), or more frequently it's a response to the anxiety and loss of control that a diagnosis brings. A good example of this is the reaction of people diagnosed with human immunodeficiency virus (HIV), which causes AIDS. Before the advent of antiretroviral therapies, rates of depression were as high as 31 per cent in people who had been told they were HIV positive.[3] The high proportion was largely caused by anxiety and the knowledge that there was no treatment for a fatal disease. After antiretroviral therapy, the rates fell dramatically. Depression is a silent partner in illnesses that remain difficult to treat. It's a logical response to the diagnosis of a horrible disease that will have an uncertain – or an all-too-certain – course.

The goal of biomedical research is remarkably simple. Provide a scientific explanation for diseases and use this information to inform diagnoses, and either come up with effective treatment or, if possible, prevent the disease from occurring in the first place. This goal has been reached for a number of diseases, and the hope is many more will fall to the scientific sword, protecting us from harm. Vaccines are the best treatment we have to prevent a disease (see Chapter 2), using the immune system to protect us. If only we could vaccinate against all diseases.

Although this goal seems straightforward, it is easily the most difficult thing for us to achieve. In retrospect, getting to the moon was easy, even though, in the famous words of John F. Kennedy, it was chosen because it was difficult. Halting climate change is, of course, a difficult challenge (see Chapter 13), but discovering a new medicine? This involves scientists of multiple specialities, depending on the disease, including immunology, biochemistry, genetics, neuroscience, molecular biology, pharmacology, bioinformatics and cell biology, to name but a few; chemists (who make the drugs); doctors (again, of many specialities) and clinical trials specialists and experts in regulatory affairs to make sure everything is done properly. Someone has to pay for all this hard work, so we need taxpayers (who pay for a lot of the research that underpins drug discovery), philanthropists and charitable foundations, financiers, venture capitalists, drug companies, people in business development and lawyers, to keep everyone in line. All in all, a highly complex ecosystem of many specialities and skillsets. It takes a city rather than a village.

Considering all this complexity, what is the actual cost of developing a new medicine from drug discovery through clinical trials to approval? Unsurprisingly, it's a lot. An awful lot. An analysis of drug development costs for 98 companies between 2003 and 2013 found that the average cost of getting a drug to market, where it can start to recoup its costs, is $2.6bn.[4] This is 14.5 times more than it was in the 1970s and 6.3 times more than the 1980s. The reasons for the increase include increased complexity in clinical trials, regulatory burdens, increased focus on areas where failure rates are high (this cost covers the impact of failures as well as successes) and increased demand from ultimate payers (e.g. insurance companies) to provide evidence of efficacy. The numbers appear to be enormous. Contrast these costs with the development cost for the first iPhone, which came in at a mere $150m.[5] So if you want to make money, perhaps go into electronics rather than pharmaceuticals.

Drug companies actually spend vast amounts of money on research and development, the aim of which is to produce new medicines. For example, Roche spends on average $8.4 billion per year on research.[6] This is 25 per cent of the *entire* budget spent by the US National Institutes of Health (NIH), the world's leading supporter of medical research. In 2018 the whole research spend of the UK's Medical Research Council came to just over $1 billion, or one-fifth of what one major drug company spends on research.[7] One problem with these comparisons is that the cost to drug companies includes clinical trials, but it remains a striking fact that drug companies spend a lot more than governments on biomedical research. Profits go back into research, although obviously there are also payments to shareholders. One fear is that should profits fall, a lot less money will go back into research to develop new medicines, although that has been disputed.[8] What is ideal, of course, is that those funded by governments to carry out research will collaborate with drug companies; increasingly, government and industry act as co-funders and work together to achieve the ultimate goal of new medicines for patients.

When we look more closely at the costs to see how activity and spend break down, we see some interesting features.[9] The drug-discovery part takes the most time. This is the phase during which companies either find a target in a disease to fire a drug at – or fire a drug at a target discovered by someone else. This phase usually involves testing the new medicine in animals, which are given a specific disease that has similar features to the disease in humans. There are so-called animal models of many diseases, including cancer, arthritis, inflammatory bowel disease and neurodegenerative diseases. The use of animals is controversial, although it is currently needed to meet government regulatory requirements for most drugs, and there is progress in replacing them with human tissues and cells, which are more likely to be predictive of ultimate efficacy in humans. This pre-clinical phase can stretch from 3 to 20 years and can range in cost from millions to billions of dollars.

Once a company has a new medicine, it applies to the FDA for a new drug application (NDA). Once the application has been provided, the company can move to the clinical trials stage. The FDA and its European equivalent, the European Medicines Agency (EMA) have to give their approval for any new drug to be tested in clinical trials and, more important, for launch onto the market in the US or Europe. The FDA was set up in 1906 to ensure that food and drugs on sale in the US are as safe as possible; partly on the back of an incident in 1902, when a vaccine for diphtheria killed 13 children in St Louis, Missouri.[10] The first goal of the FDA is to check for 'adulteration' of products on sale in the US, particularly after 1937 when an antibiotic called Elixir Sulfanilamide killed over a hundred people because of its formulation in a toxic untested solvent. This event led to the FDA being given powers to designate drugs as safe and to be used only under the supervision of a medical professional. There is often tension between the FDA and drug companies and patients, with claims that the FDA regulations unacceptably lengthen the time it takes to bring a drug to market. Concerns about this issue came to a head during the AIDS epidemic, with HIV activist organisations putting pressure on the FDA to speed up the process. In response to these criticisms, the FDA introduced its 'Critical Path Initiative', which has helped matters somewhat.

The FDA mandates three phases of clinical trials. This is a stepping-up process, which is considered the safest way to do things. Phase 1 is only about safety – the experimental drug is given to healthy people and any 'adverse events' are closely monitored. An adverse event could be headaches or nausea or more serious changes in liver enzymes, which indicate that the new medicine might harm the liver (where toxins are metabolised in our bodies). Phase 2 is the first time the drug is tested in patients with the disease, and usually involves tens of patients and tens of controls. The gold standard is a double-blind placebo-controlled trial. It is called 'double-blind' because neither patient nor doctor knows who is on the drug or the placebo. Finally, Phase 3 is a repeat of Phase 2 but consists of a lot more patients, with

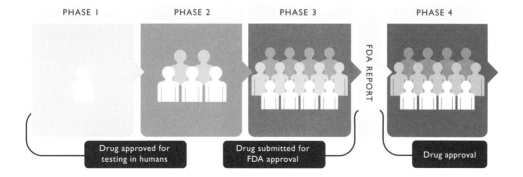

sometimes thousands in the treated group and thousands in the control group. If all is well with Phase 3, the FDA will then approve the drug for the market. Finally, post-launch, the company will observe what's happening in the field with many patients now on the new medicine. This is known as Phase 4. A new medicine can fail at any phase of this complex and time-consuming process.

THE FDA MANDATES FOUR PHASES FOR THE TESTING OF A NEW DRUG. EACH PHASE INVOLVES MORE PATIENTS UNTIL APPROVAL FOR USE AFTER PHASE 3, ONCE EFFICACY AND SAFETY HAVE BEEN ASSURED. PHASE 4 MONITORS THE DRUG AFTER ITS LAUNCH. THE PROCESS CAN TAKE OVER 10 YEARS TO COMPLETE.

There are costs all along the way.[11] Depending on the disease, a Phase 2 trial can cost $50 million or more. Phase 3 can cost hundreds of millions. The FDA itself charges the company up to $2m for an NDA application, and that's before any patients have seen the drug. The number of drugs that actually make it through the punishing application process is minuscule.[12] It takes around 12 years to get from the research lab to the patient, and just 5–10 per cent of drugs tested make it to patients. Each drug company will have hundreds of drugs in development. It can only recoup the losses incurred by failure through the profits from drugs that make it.

A recent detailed analysis of success rates makes for sobering reading. The industry uses the term 'likelihood of approval' (LOA).[13] From a Phase 1 trial, the LOA for drugs tested stood at 9.6 per cent. You need nerves of steel to be in the drug discovery business. Only 63.2 per cent of drugs made it from Phase 1 to Phase 2. Phase 1 is only a safety trial and so is more likely to succeed than other phases, as safety will have usually been tested in animals (including monkeys) first. Once you get to Phase 2, 30.7 per cent make it to Phase 3. Phase 2 is when patients are first given the drug and, the drug can fail at that stage for a whole host of reasons – the usual one being a lack of sufficient efficacy, often because the overall hypothesis (that a target was critical for the disease) was wrong or not compelling enough. Or sometimes the dose can be wrong, or the patient didn't respond. Sometimes a business decision prevails – the commercial case for the new medicine may not be compelling enough (possibly because rivals have got there first) and the company might have cold feet because of the substantial cost of Phase 3 testing. There is a 58.1 per cent chance that a drug will make it through Phase 3. If it does, there is an 85 per cent chance that the FDA will approve it and, at last, it makes it to the market. We finally arrive at an overall LOA of 9.6 per cent, or around a one-in-ten chance. Would you be happy to wager millions (which is what it costs to get to a Phase 1) on a horse running in a race with odds of 10–1? Drug companies will be backing many horses at those odds, and so there is an increased chance that at least one will romp home. This also applies to venture capitalists who often provide the investment for early-stage biotech companies. And we see successes emerge from this game of substantial attrition.

The question then becomes, how do drug companies shorten these odds? Science is leading the way. In terms of chances of success, an interesting study examined the relative role of industry and academia in the drug-discovery process.[14] In 1980 the Bayh-Dole Act allowed universities to patent and profit from successes achieved via government-funded research. This meant that universities could make money – usually from licensing agreements with drug companies and subsequent royalty payments – should any drugs make it to market. This incentivised universities to carry out drug-discovery research, which then became increasingly necessary for raising income to support academic activities when governments cut university funding. Between

1998 and 2007, of the 252 new drugs approved by the FDA, 191 came from drug companies and 61 came from universities or small biotech companies, which were usually set up by academics. Academics often bring new ideas and concepts to big drug companies and the relationship between academia and industry continues to grow and develop. Academics are helping to lower the odds of new drugs actually making it, although there is no evidence as yet that the targets they find or drugs they develop have any more chance of making it. A recent study, however, has shown that the overall success rate has risen to 14 per cent.[15]

The odds can also be shortened by a new approach called 'precision medicine'. Can you precisely identify which patients to treat for a given disease, given their precise symptoms? This might even come down to precise differences in genes or proteins in their bodies, even before symptoms manifest. Finding out what actually causes different diseases is challenging. For some diseases now considered the 'low-hanging fruit' of the drug-discovery business, it was possible to pinpoint what was wrong for a given condition and then correct it accordingly. A simple example is infectious diseases that were shown to be caused by bacteria. Answer? Antibiotics to kill the bacteria. Another example is Type 1 diabetes, caused by a deficiency in insulin. Answer? Replace the insulin. These discoveries have saved many millions of lives.

Yet when it comes to treating more complex diseases, it is useful to have a 'biomarker'. This is a change in the body during the disease process, which, if corrected, will predict that the drug will work. These are especially useful for certain types of cancer. A good example here is HER2-positive breast cancer.[16] If breast tumours are shown to carry a biomarker called HER2, they can then be treated with a drug that targets HER2. One in five women with breast cancer are HER2 positive (meaning that HER2 can be detected on their tumours). When given Herceptin, HER2-positive patients will see an extension in the time it takes for the disease to progress, which, overall, can extend the life of a patient. Importantly, it only works in HER2-positive

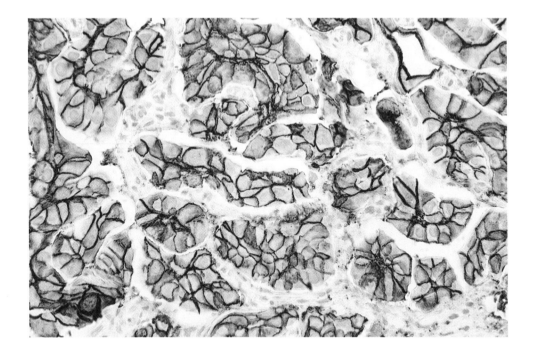

TISSUE FROM A PATIENT WITH BREAST CANCER. THE PROTEIN HER2 IS STAINED BROWN. A POSITIVE TEST MEANS THAT THIS PATIENT IS LIKELY TO RESPOND TO A DRUG THAT TARGETS HER2.

breast cancer patients. A recent analysis has shown that around 5 per cent of trials currently incorporate biomarkers.[17] If they are used, the chance of success increases, with an LOA rising to 25.9 per cent, meaning that the odds are shortened from 10–1 to 4–1 – still not a home run, but heading in the right direction nonetheless. Work on biomarkers proceeds apace and will continue to reduce the odds even more.

All that effort eventually delivers what we all want – a new medicine that works on a particular disease. Many of these new medicines will emerge in the coming years. But given the considerable effort and development costs, how much can a drug company charge for a new treatment?

A good way to illustrate the tensions that are emerging regarding the costs of new treatments and who will pay, and what is likely to happen in the future, is to look at specific diseases. Leber's congenital amaurosis (LCA) is a rare genetic disease that causes blindness in children. It afflicts 1 in 40,000 children, and 18 genes have been implicated in the disease.[18] These genes are defective, and this results in the abnormal development of specialised cells in the retina of the eye, called photoreceptors, which detect light. One type of LCA involves a gene termed RPE65, and in 2017 the FDA approved gene therapy for this type of LCA.[19] Efforts began back in 2007, which illustrates the time it takes for treatment to eventually make it to the clinic. The therapy, however, worked, preventing blindness in children with this devastating disease. This treatment represents a ground-breaking approach to treating genetic diseases and is seen as a predictor of what is to come for many conditions that are mainly genetic in nature. The company that makes the treatment, which has the trade name Luxturna, is called Spark Therapeutics, and the list price for the treatment is $850,000 – or $425,000 per eye. Will health insurance companies or government health agencies cover the cost, or will it fall to the parents of the child who is going blind? The company has proposed strategies for how people might pay this enormous amount of money. Spark has offered the options of stretching the payment over several years or refunding the payment if the treatment fails. However, the Institute for Clinical and Economic Review (ICER), a US-based organisation which examines the price of medicines, said that the cost should be 50–75 per cent less.[20] ICER examined the justification of the cost by assuming a 10-20-year benefit of treatment and comparing that to the cost to the health system and wider economy if the patient were left untreated. Agencies such as ICER and the National Institute for Health and Care Excellence (NICE) in the UK are there to defend patients and ensure prices are justified. Spark is negotiating the price with the US government and private health insurers. In the case of Spark and Luxturna, the treatment is a big breakthrough for patients. These discussions are hopefully a harbinger of what's to come, and the hope is that something can be worked out to allow patients to access the new treatment.

A second example of the challenges of setting a price for a treatment can be seen in the approach to cystic fibrosis (CF). Recently there has been

HEALTHY LUNGS

LUNGS AFFECTED BY CYSTIC FIBROSIS

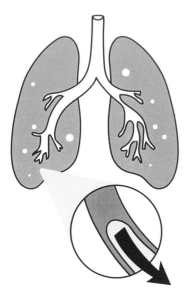

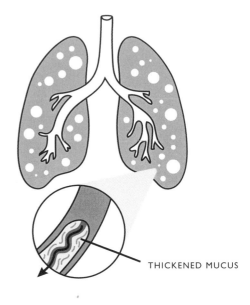

THICKENED MUCUS

CLEAR AIRWAY

AIRWAY WITH CF

CYSTIC FIBROSIS. A SALT IMBALANCE IN THE LUNGS LEADS TO THICKENED MUCUS, WHICH PROMOTES BACTERIAL GROWTH, LEADING TO INFLAMMATION AND LUNG DAMAGE. THE LATEST TREATMENTS RESTORE THE SALT BALANCE.

remarkable progress in the effort to find new treatments. CF is a genetic disease that affects mainly the lungs, but also the pancreas, liver, kidneys and intestine. It afflicts 1 in 3,000 newborns. Long-term health issues include difficulty breathing and frequent lung infections, which eventually lead to lung failure and death. It is caused by mutations in the gene for a protein called cystic fibrosis transmembrane conductance regulator (CTFR). CTFR does its job on the surface of cells in the lungs, regulating the salt balance outside the cell. The mutant protein is unable to do this, which means that the fluid in the lungs becomes thicker. This fluid puts pressure on the lungs and also encourages bacteria to grow, which causes inflammation. People who have one copy of the mutated gene and one copy of the normal gene (remember, we have two copies of all genes: one from our mother and one from our father) are carriers: but if both copies of the gene they carry are mutated, they will develop CF, because the resulting mutant protein can't do its job properly.

Because the disease is genetic, there has always been a hope that a gene-therapy approach might work. Is it possible to replace the faulty gene with one that works? One option is a lung transplant, although this is problematic because of a lack of suitable donors. However, the US drug company Vertex Pharmaceuticals discovered a new type of medicine, Ivacaftor, approved for use in 2012, which they developed in conjunction with the Cystic Fibrosis Foundation (CFF), a non-profit organisation that aims to improve the quality of life for people with CF.[21] Ivacaftor works in people with one particular set of mutations. It binds to the mutant protein and enhances its function; it basically encourages the mutant protein to do its job. Ivacaftor showed significant efficacy in clinical trials and was approved by the FDA. The cost of Ivacaftor is $311,000 per year, which is in the same range as other diseases that are as rare as CF.[22] An editorial in the *Journal of the American Medical Association* called the price 'exorbitant'.[23] It mentioned that a charitable foundation had supported the discovery of the drug, and how publicly funded research at the National Institutes of Health had provided important early insights into CTFR and CF. Vertex came back with a statement saying that it took 14 years of research, mostly funded by the company, before the drug won approval.[24] The CFF had provided $150m in funding in exchange for royalty rights. It sold these rights for $3.3 billion, which it is spending on further research.[25] Vertex also said it would provide the drug for free for people in the US with no insurance and a household income of under $150,000. In the UK, an assessment was done on what is called the quality adjusted life year (QALY), which measures life expectancy and the burden a disease has on a sufferer's life. One QALY equates to one year in perfect health. The score can range from 1 (perfect health) to 0 (dead). For someone with cystic fibrosis, this number will fall as the years go by and finally, the patient will sadly die of the disease.

A new drug has to impact on this, either slowing down this decline (which is what Ivacaftor has been shown to do) or stopping the decline altogether (which is a cure for the disease). The study concluded that the high price was justified because of the benefit the treatment brings to patients.[26]

A second Vertex drug, called Orkambi,[27] was then approved. This is a combination of Ivacaftor and another drug called Lumacaftor. Orkambi has been shown to be effective in patients with the much more common F508D mutant, which occurs in as many as 80 per cent of CF patients in Ireland. People with this mutation differ from those with the mutations that respond to Ivafactor alone. In their case, the CTFR protein is broken in two ways, not just one. It can't do its job correctly on the surface of the cell, but it also never even gets there – it stays inside the cell. Lumacaftor kicks the mutant CTFR onto the cell surface, and then Ivacaftor does its job and makes the mutant protein work properly. A double punch. Orkambi is priced at $259,000 a year in the US.[28]

Finally, a third medicine was approved, termed Trikafta. This is a combination of three drugs: Ivacaftor, Elexacaftor and Tezacaftor.[29] It has been shown to work five times better than Orkambi (now we're talking) and has a list price in the US of $311,000 per year.[30] The Irish government recently announced that it had reached an agreement with Vertex to reimburse the cost of the drug for Irish CF patients.[31] Previously, Vertex provided Orkambi to patients at a total cost to Ireland of €400m. The charitable foundation CF Ireland has expressed great excitement at the news. About 800 patients will be eligible for treatment and will greatly benefit from Trikafta. The vista for the disease CF is therefore likely to improve beyond all recognition.

But CF is a rare disease. What happens if a new and highly effective treatment for a more common disease emerges? For a new medicine, pricing begins with the drug company estimating its value, which is based on what they 'think the market will bear'. A blockbuster first-in-class treatment (which means a brand-new drug for a major disease, not a lookalike of something that's already out there) can command a high price. An excellent example of this is Solvadi, which is made by the drug company Gilead. This drug is a cure for hepatitis C, a dangerous virus that infects the liver and irreversibly damages it as well as increasing the chances of liver cancer.[32] Worldwide 150 million people are infected with hepatitis C and it causes 400,000 deaths per year.

Gilead developed a drug to kill the virus, which worked spectacularly in clinical trials. Once the FDA had approved it, Gilead set the price at $1,000 per tablet or $84,000 for a complete treatment course.[33] The price has since come down because of competition in the marketplace, which is always a good thing. Rival company AbbVie launched a competitor and set the price at $26,400. More recently, the Egyptian drug company Pharco Pharmaceuticals carried out a trial on a combination of Gilead's drug with another drug called Ravidasvir.[34] This was shown to cure 97 per cent of patients in a Phase 2–3 trial, which is a higher rate than when using Solvadi alone. Even though Gilead first made Solvadi, the projected cost of this combination has been set by Pharco at $300 per patient in total.

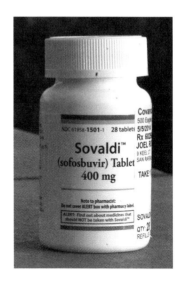

THIS DRUG CURES PEOPLE OF THE VIRAL DISEASE HEPATITIS C. IT CAME AT A HEFTY PRICE WHEN IT WAS LAUNCHED BUT COMPETITION HAS REDUCED THIS CONSIDERABLY.

Egypt is actually a good case in point, as it has the highest incidence of hepatitis C in the world. The Egyptian Minister for Health, Adel El-Awadi, has said that if Egypt were to treat all its hepatitis C patients with Gilead's drug at the list price, it would absorb the country's entire health budget. To Gilead's credit, it then negotiated a price for Egypt that was 99 per cent cheaper, or 1 per cent of the list price. Overall, the current prices for hepatitis C medicines mean that only 3 million people are being treated,[35] but this will change as rivals enter the marketplace, eventually resulting in many millions of lives being saved.

Gilead are also the leading company developing an anti-viral drug for SARS-CoV-2, the virus that causes COVID-19. Their drug Remdesivir was shown to be effective against Ebola virus, which is related to SARS-CoV-2 (both are RNA viruses) and has the potential to treat COVID-19.[36] Exactly how Remdesivir will be priced is unknown, but the CEO of Gilead, Daniel O'Day, has said it will be 'affordable'. When it comes to a vaccine for COVID-19, pharmaceutical company Johnson & Johnson, one of the leading companies developing a vaccine, have said they will make the vaccine available for free. This would be unprecedented, and is in part due to the fact that the US government is providing half the funding to develop the vaccine.

Once a drug company sets the list price, it can be bargained over – a bit like haggling with a market vendor.[37] Rebates can be applied to employer-sponsored schemes or self-insured health plans or indeed to governments who are covering the costs (like the HSE in Ireland). Pharmacies also get involved. Price is subject to negotiation at multiple levels. Constant monitoring is also needed. Whatever about new drugs arriving to the market, drug companies can also raise the price of drugs already on the market.

An egregious example of price-gouging happened when a company called Mylan raised the price of its EpiPen by 500 per cent, from just under $100 to more than $600.[38] The EpiPen is used to save the lives of people who are having a major allergic reaction. In another example, from 2002 to 2013, the price of insulin (another life-saving medicine) increased threefold.[39] And from 2012 to 2019, the price of AbbVie's rheumatoid arthritis drug, Humira, which blocks an inflammatory protein called TNF in rheumatoid arthritis, was increased from $19,000 to $60,000.[40] The current reality is that the price of branded medicines such as these are growing at a rate that far exceeds inflation.

A recent survey revealed some interesting differences between Ireland and other countries:[41] Irish patients are paying over six times the international average for generic drugs. A generic drug is a drug that contains the exact same chemical as a drug that was originally protected by a patent. Generic drugs are allowed once the patent on a drug has expired. Drug patents last for 20 years, although that includes the time it takes to get the drug to the clinic

AN EPIPEN PIÑATA FILLED WITH GOLD COINS TO REPRESENT THE GREED OF THE PHARMACEUTICAL INDUSTRY IS SMASHED BY ACTIVISTS. MYLAN INCREASED THE PRICE OF THE EPIPEN BY 500 PER CENT TO $600, BUT THEY WERE SUBSEQUENTLY FORCED TO LOWER THIS.

once it has been patented, which takes on average eight years. This means on average 12 years of patent protection when the drug company can set the price to recoup expenditure and make a profit. Generic drugs in Ireland are more expensive than they are in most other European countries; yet Ireland has the lowest price of the 50 countries assessed for the erectile dysfunction drug Viagra (could this be to encourage Irish men to use it more?) and is also cheapest for the cholesterol-lowering drug Lipitor, which protects against heart attacks. Overall, Ireland ranked 16th in the world for drug prices.

When it came to branded drugs, Ireland was slightly cheaper than average. The Health Service Executive in Ireland pays for branded medicines by joining forces with 14 other countries and negotiating the price with each drug company. So we aren't doing too badly when it comes to drug pricing, and the HSE tries to ensure everyone gets access to the treatments they need. One issue is access to new drugs, with the Irish government being accused of delays in getting approval and agreements on prices.[42] The total drug bill for Ireland has risen to €2.5 billion in 2018, and it is a growing concern.[43] It is imperative to keep drug prices down so that patients get access to new drugs and treatments.

What does the future look like for new drugs and their pricing? For much of the last decade, the FDA approved around 20–25 new products per year. In 2018 the FDA approved 59 new drugs and it's likely that this level of approval will continue into the future.[44] The new drugs include brand-new treatments for migraine, a wide range of cancers (including several types of leukaemia), pain in endometriosis, nausea for people on chemotherapy, and severe epilepsy. Important new approvals in the past few years include treatments for cancer that involve mobilising the immune system against tumours and gene therapy approaches for different rare diseases. Again, all these drugs are expensive. Novartis recently announced a new gene therapy, Zolgensma, for a paralysing disease called spinal muscular atrophy. This showed remarkable efficacy in clinical trials. Novartis has said it will price this one-time therapy at $2.1m.[45] A reason for this high cost is that few people have this disease, so the cost of development must be spread over fewer patients. It is currently the most expensive treatment ever.

All these new medicines benefit patients, but all are coming at a high price. If the HSE spent a vast amount on very few patients with severe diseases, would that compromise their spending on other areas where a lot more people might benefit? Doctors always have to make decisions when they are treating their patients and this is somewhat similar, especially if there are limitations. During the COVID-19 pandemic, doctors sometimes had to decide who to put on a ventilator, as there were insufficient numbers of ventilators available. The HSE has to decide how to spend its budget, which will become an increasing challenge in the face of new expensive medicines. The pot of money available is not infinite, and there have to be trade-offs with the cost to a health service if a patient doesn't get the new treatment, which, over the life of the patient, can be substantial. This provides another justification for the high cost. Perhaps a core list of minimum medicines needed for a basic healthcare system can be used, which lists the most efficacious, safe and cost-effective medicines for priority conditions.[46] New medicines can then be evaluated and added to that list.

Whatever way you look at it, there are exciting and challenging times ahead for all of us. Tough decisions will have to be made, but ultimately all that medical research is giving rise to new treatments from which everyone can benefit, irrespective of their ability to pay. How likely is that, I hear you ask? Isn't that the question that has dogged healthcare, both in the developed and developing world, for decades? **The bottom line: let's hope we find a way forward when it comes to making new drugs available for those who need them, regardless of whether they can pay or not.**

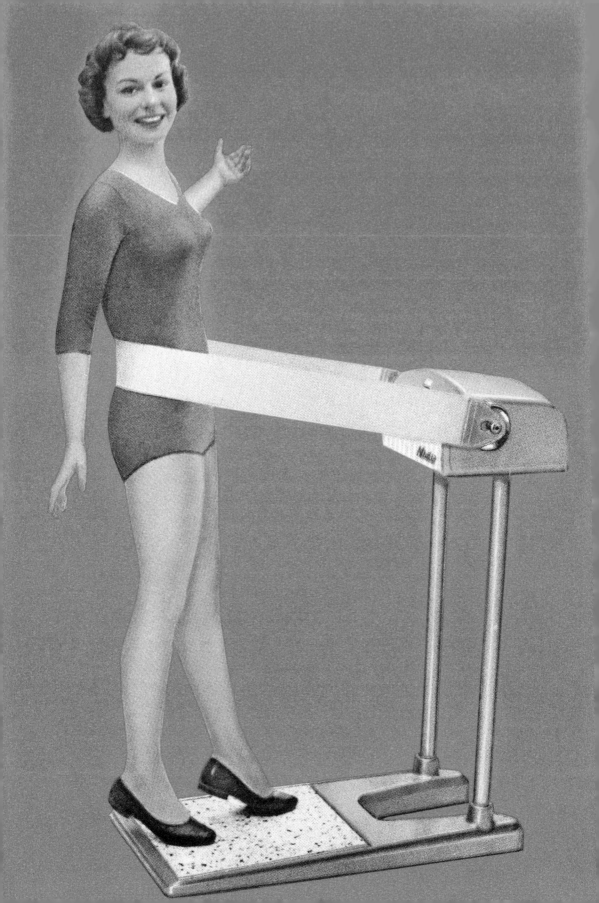

CHAPTER 4

WHY DO YOU BELIEVE IN DIETS?

—

'I tried every diet in the book. I tried some
that weren't in the book. I tried eating the book.
It tasted better than most of the diets.'

—

Dolly Parton

HAVE NEVER GONE on a diet. Now there's an admission! I'm slightly overweight (note the *slightly*), but then lots of us are. Weight Watchers Ireland, an organisation that helps people to lose weight, has 100,000 members.[1] Ninety-five per cent of them are women. Almost half of Irish women are overweight, and the vast majority want to lose weight. The international slimming industry makes billions of euros each year, yet there is little if any scientific evidence that any of the diets they tout work in the long term, or work better than simply reducing calorie intake. For Irish men, the situation is worse. Sixty-six per cent are overweight, like me.[2] Even more worryingly, one in four Irish children are overweight at a time of life when they should be running around burning calories.[3] Ireland has one of the highest rates of obesity in Europe, with one in four adults now classified as obese.[4] Globally, people are becoming overweight or obese. Since 1975, worldwide obesity has almost tripled. In 2016, when a detailed assessment was last done, 1.9 billion adults were overweight and, of these, 650 million were obese.[5] All that fat, and all that dieting. It's a big problem. What's going wrong and what can be done about it?

First, we should define what it means to be overweight or obese. Nutritionists use body mass index (BMI) as a way to define these terms. The idea of BMI was first suggested by Belgian sociologist and statistician Adolphe Quetelet.[6] He was interested in a way of defining the average for many

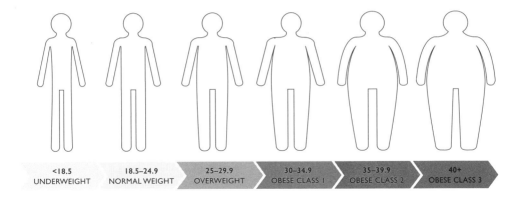

| <18.5 UNDERWEIGHT | 18.5–24.9 NORMAL WEIGHT | 25–29.9 OVERWEIGHT | 30–34.9 OBESE CLASS 1 | 35–39.9 OBESE CLASS 2 | 40+ OBESE CLASS 3 |

different human traits since it was well known that humans come in all shapes and sizes. He devised the BMI (which was originally called the Quetelet Index) as a way to measure a person's body mass by taking into account their height and weight. (Using weight alone didn't suffice, as someone who is tall would have a bigger weight than someone who is small, and yet they wouldn't necessarily be overweight.) The term BMI was coined in 1972 by American physiologist

BMI IS CALCULATED BY DIVIDING A PERSON'S WEIGHT IN KILOGRAMS BY THE SQUARE OF THEIR HEIGHT IN METRES. IF YOUR BMI IS OVER 30 YOU HAVE A HIGHER RISK OF DYING FROM COVID-19.

Ancel Keys, who, in response to attacks from professional rivals (not unusual in the world of science), said, 'if not fully satisfactory, it's at least as good as any other relative weight index, as an indicator of relative obesity'. BMI is a person's weight in kilograms divided by the square of their height in metres.[7] If your BMI is less than 18.5, you are defined as underweight. If you are in the range 18.5–25, you are of normal weight. If your BMI falls between 25 and 30 you are overweight and if you are over 30 you're in (literally) big trouble as you are obese. There are three classifications of obese: Class 1 is a BMI between 30 and 35, Class 2 is 35 to 40 and Class 3 is 40 or higher. A second measure that's used is waist circumference, which can be as useful as BMI to assess whether someone is overweight or obese. The reason why measuring BMI is important is because overweight and obese people are at a much higher risk of developing a large number of diseases.[8] These include coronary artery disease, Type 2 diabetes, high blood pressure, osteoarthritis, stroke, depression, at least ten cancers, including breast and colon cancer, and, most recently, COVID-19. Being overweight or obese also substantially increases the overall risk of mortality. In a 2009 study of 900,000 people, those who were overweight or underweight had a mortality rate higher than people of normal weight as defined by BMI.[9] The main causes of death were heart attacks, stroke and cancer. Overall, overweight people were around 50 per cent more likely to die of these diseases than people of normal weight. Being overweight has now overtaken smoking as the leading preventable cause of death.[10]

Another, more insidious, aspect of being overweight concerns body image. For girls and women, and increasingly in boys and young men, there is huge pressure to be slim. This wasn't always the case. The history of how humans view body image is a fascinating one. In Ancient Egypt, the perfect figure for women was slender, with narrow shoulders. In Ancient Greece, the male body was seen as more aesthetically pleasing, and women were inclined not to worry about their own appearance so much, and so were more inclined to be slightly overweight. During the Italian Renaissance, a fuller figure in women was much preferred as it was a sign of fertility and affluence. In Victorian times, a narrow waist was seen as attractive, with ampleness encouraged in other parts of a woman's body, giving rise to the ridiculous situation where women had to squeeze themselves into whalebone corsets. In the 1920s a more androgynous look in women became desirable. After World War II another shift happened, with a fuller figure again in vogue: in fact, in the 1950s women would sometimes consume weight-gain supplements to achieve what was considered a desirable figure. From the 1960s onwards the style began to shift inexorably towards thinness. Cosmetic reduction surgery became increasingly popular, as did dieting and aggressive exercise. Those trends continue and have, if anything, become more punishing for women – who have yet to escape the tyranny of being 'evaluated and oppressed by their appearance' as one commentator has put it.[11]

Physical appearance continues to be seen as a critical facet for girls and women. In a recent survey carried out by Unilever (which own the brand Dove, so this survey must be taken with a pinch of bath salts), only 4 per cent of women think they are beautiful.[12] Another study has shown that 91 per cent of women are mostly unhappy with their bodies while 40 per cent have seriously considered cosmetic surgery to correct their perceived flaws.[13] Ninety-seven per cent say they have at least one negative thought about their body image every day. And social media is only making things worse, exacerbating the natural human tendency to compare ourselves to others. The outcome of all of this is serious. It is in part responsible for the epidemic of anxiety and depression in our teenagers, but more importantly, all of this increases the risk of serious eating disorders such as anorexia and bulimia. The problem is less common in men, although not insubstantial, with 20–40 per cent expressing dissatisfaction with their bodies.[14]

The media needs to change the way it portrays women, and this has been a long-term goal for advocacy groups trying to help teenage girls.[15] Encouraging the self-esteem of girls is also key to progress, but these goals are proving difficult to achieve. Being overweight or obese can be a serious issue, both medically and psychologically. On his *Late Late Show*, James Corden recently highlighted the psychological impact of his struggle with weight. He said that despite trying very hard he had never been able to control his weight, saying that he had 'good days and bad months'. James was speaking out against another commentator in the US, Bill Maher, who had said that 'fat-shaming' needed to make a comeback. Maher lambasted overweight people, saying they lacked self-control. Corden branded this as bullying and said that the one thing we know about bullying is that it never works; it just makes people feel bad about themselves. Corden was referring to the stigma of obesity. A multi-disciplinary group of international experts recently issued a consensus statement in an effort to end the stigma of obesity. Its stated aim is to 'encourage education about weight stigma to facilitate a new public narrative about obesity, coherent with modern scientific knowledge'.[16] This is another step in the right direction towards ending 'fat-shaming'.

So what causes people to be overweight or obese? There are the obvious reasons: excessive food intake and lack of exercise. If we start with excessive food intake, there are reliable recommendations of how many calories people should consume per day. We need to consume a minimum amount of food per day to sustain us. Food provides us with energy to build muscle and maintain bodies and to keep our brain and other organs running. Nutritionists divide what we eat into three main food groups: carbohydrates (like sugar), fats (like butter) and proteins (which are found in plants and animal meat). We also need to take in vitamins and minerals as they are needed for various things in our bodies to work properly, including helping our blood clot when we injure ourselves (vitamin K) or to help with bone strength (vitamin D) or to help enzymes to work properly, for example, during digestion (B vitamins).

The energy part of food is used as a measure for how much food we should eat. The International System of Units (the modern form of the metric system) measures energy in joules. Some countries display food energy in kilojoules (kJ), but a different measurement is also used, termed calories (also rather confusingly called kilocalories (kcal)), which is an older unit of measurement. In the European Union both kilojoules and kilocalories are displayed on foodstuffs. In Canada and the US, only 'calorie' is used. A calorie is defined as the amount of heat energy needed to raise the temperature of one kilogram of water by 1 degree centigrade. The number of calories in a given amount of food can be calculated by measuring the amount of carbohydrates, fats and proteins in the food and since we know how many calories are in each of these by weight, the total calories can be figured out. A device called a bomb calorimeter can also be used, where the food is burned and then measuring how that changes the temperature of water surrounding the calorimeter.

Fats and alcohol have the largest amounts of calories per gram at 8.8 calories and 6.9 calories each respectively. Protein and carbohydrates have about four calories per gram. Many governments require food manufacturers to label the energy content of their products. Recommending how

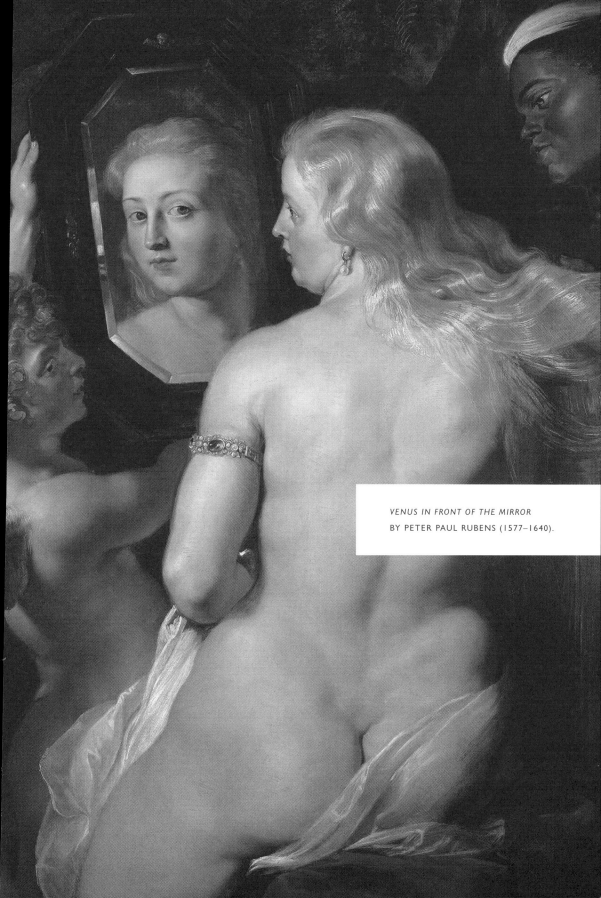

VENUS IN FRONT OF THE MIRROR
BY PETER PAUL RUBENS (1577–1640).

many calories to eat per day is complicated, as it varies based on age and level of exercise. For example, current recommendations as to an optimum calorie intake per day are 2,500 calories and 2,000 calories for men and women respectively,[17] when they are aged between 31 and 50 and have a rate of exercise of walking 2–6 km per day at a pace of 5–6 km per hour – see, I told you it was complicated. Similar calculations can be done for people of different ages and levels of exercise. But those numbers for the 31–35 age group are seen as average overall. Our brains use 20 per cent of the food we eat – all the neurons firing away make them excessively greedy – with other organs in our bodies using the rest. If you don't follow these recommendations you are at risk of becoming malnourished (which will eventually kill you) or at the other extreme, becoming overweight or obese.

In addition to a lack of exercise and excessive food intake, there is a third reason for obesity where the evidence continues to grow: genetic susceptibility.[18] It turns out that there is a highly reputable and growing body of evidence, which points to our genes as a significant reason for obesity. Studies on identical twins have revealed that obesity is 70–80 per cent genetic.[19] If you carry variants of genes tied into obesity you will have a much harder time controlling your weight. The genetic contribution is as high as that for the genetic basis for height and substantially higher than for many diseases where genes have been implicated. The increased prevalence of obesity in recent times is likely to be because some of us have a propensity to become obese because of our genetics, combined with a modern lifestyle (such as the easy availability of high-calorie food), which might lead us to overeat or take less exercise. So many people who are obese carry a double burden: as well as having an inherent risk of becoming obese because of their genes, they are also considered to lack willpower and be lazy.

One of the earliest observations that obesity might be genetic was carried out in 1997 by Irish doctor Stephen O'Rahilly, who was studying a massively obese four-year-old boy.[20] Given the boy's family history, O'Rahilly figured his obesity could be genetic. When the boy was fed a single test meal of 1,125

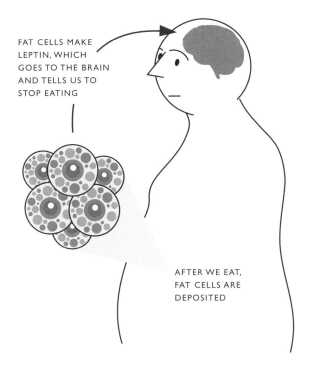

FAT CELLS MAKE
LEPTIN, WHICH
GOES TO THE BRAIN
AND TELLS US TO
STOP EATING

AFTER WE EAT,
FAT CELLS ARE
DEPOSITED

IRISH PHYSICIAN
STEPHEN O'RAHILLY
DISCOVERED THAT
A MUTATION IN THE
LEPTIN GENE LEADS
TO DEFICIENCY THAT
CAUSES OBESITY
AS A RESULT OF AN
UNREGULATED APPETITE.

calories – half the recommended *daily* intake for an adult – he asked for more food. He was found to have a defect in a gene for a protein called leptin – his body couldn't make this hormone. When he was given an injection of leptin, his appetite went back to normal, and unlike Dickens' Oliver Twist he no longer asked for more. This important study supported the idea that becoming obese could have a genetic basis. In one fell swoop, it destroys Maher's assertion that food intake is under voluntary control. In that boy's case, it all came down to leptin levels, with less leptin meaning more food being eaten. Leptin is made by fat cells in your body. When we eat fat, the fat cells release leptin, which goes to the brain and tells us to stop eating. When you lose weight, the number of fat cells in your body decreases. After you eat, you will make less leptin and your appetite goes up and you will then eat more. It's a bit like a thermostat, except it's a fat-o-stat. More fat, more leptin, eat less. Less fat, less leptin, eat more. The leptin system probably evolved to stop us becoming too thin, which can be dangerous, or too overweight, which might make us more susceptible to predators because of reduced mobility or put us at greater risk of the diseases listed above.

A genetic deficiency in leptin is a rare cause of obesity, but differences in other genes that make proteins tied into appetite are more common. A host of other genes that lead to obesity have since been found since O'Rahilly's initial discovery.[21] These are all tied into appetite and feeling of satiety, or feeling full. These include the receptor for leptin (which senses leptin in the brain). If that is broken or missing because of a genetic change, your body won't respond to your own leptin. People who are obese for that reason won't lose weight if you inject them with leptin because they can't respond. The other gene variants tied into obesity include genes for proteins called MC4R (which accounts for 5 per cent of people who are obese), FTO, ADIPOQ, PCSK1 and PPAR-gamma. At least 50 genetic variants have been shown to associate with risk of obesity and it is likely that people who are obese will have different combinations of these and other genes, with each playing a minor role but when combined increase the risk. More work needs to be done in this area, but ultimately it might be possible to test people for genetic risk, and then depending on which genes are defective, perhaps correct the defect and then reduce the risk of obesity. But those days are still some way off.

So if you are overweight or obese, or feel you are at risk of either, what can you do about it? Surgical procedures are available, but these are usually only suitable for people who have long-standing obesity and who have not responded to lifestyle approaches, including behavioural intervention and dietary change. The commonest surgical intervention is called bariatric surgery, which reduces the size of the stomach by actually removing part of it. This is only approved for people with a BMI of at least 40, although research is emerging that indicates it may be useful for people with BMIs of above 30 with other diseases linked to obesity, such as Type 2 diabetes, sleep apnea or high blood pressure.[22] It seems rather extreme but can be a lifesaver in the morbidly obese. Bariatric surgery can cause remission of Type 2 diabetes in up to 80 per cent of cases and also cure sleep apnea for many.

Certain drugs have also been shown to bring benefits.[23] Orlistat decreases absorption of fat from the gut. Phentermine increases norepinephrine release, which decreases food intake. Lorcaserin can decrease appetite by acting on 5-HT2C receptors in the part of the brain that controls appetite. Some drugs used to treat Type 2 diabetes, such as Metformin and Liraglutide, have also been shown to affect appetite and decrease body weight. New possible treatments are in development and may well involve the judicious use of some of these medicines (which can have side effects). Some of the drugs that are in development for managing obesity are delivering results in the early clinical trial phase that are equivalent to the results seen following bariatric surgery.

But most people start their weight-loss journey with a diet. Dieting is defined as the practice of eating food in a regulated manner to affect body weight. This can mean attempts to put on weight but much more commonly it can mean attempting to lose weight. Dieting is a complex and fraught business and has its own qualified professionals, called dieticians. Their role is crucial, as many people need help with nutrition, including older people who might be malnourished, or neonatal babies or people with a whole host of diseases who might need nutritional supplements or even full nutrition if they can't feed themselves. Dieticians also have an important role in helping people with eating disorders.

One of the first dieticians was George Cheyne, an obese English doctor who modified his diet to lose weight, eating only milk and vegetables, and who wrote about this diet in an essay published in 1724.[24] The first popular diet, 'The Banting', was created by William Banting who was, perhaps appropriately enough, an undertaker. He wrote a pamphlet called *Letter on Corpulence, Addressed to the Public*, which recommended a diet involving four meals a day consisting of meat, green vegetables, fruit and a glass of dry wine.[25] He advocated omitting sugar, beer, milk and butter. The booklet

AN ANTI-BANTING-ITE.

on 'The Banting' remained in print until 2007. The craze to count calories began in 1918, in a book entitled *Diet and Health: With Key to the Calories* by American doctor Lulu Hunt Peters.[26] There then followed a deluge of diets, especially from the 1960s onwards. In the US an estimated 45 million people are on a diet at any given time, spending $33 billion every year on weight-loss products – that's $33 billion a year, while 800 million people in the world are starving.[27] We are a curious species, are we not?

Diets are many and varied. In the 1970s there were all kinds of fad diets. The grapefruit diet recommended eating half a grapefruit with every meal. The Slim-Fast diet involved drinking a protein milkshake for breakfast and another for lunch. The Scarsdale diet involved eating only 700 calories per day for two weeks – it was invented by Dr Herman Tarnower who sadly was shot dead by his girlfriend (the two are not necessarily connected). Newer diets came along in the 1980s. The cabbage soup diet involved drinking two bowls of cabbage soup each day, along with bananas and skimmed milk. Smelly. The Beverly Hills diet was on *The New York Times* Best Seller list for 30 weeks and involved eating one type of food for ten days and then another for another ten days and so on. Fruit in week one, carbohydrates in week two, protein in week three: difficult to carry out in practice. Then in 1988 came the liquid diet. This doesn't mean drinking just alcoholic beverages. Oprah Winfrey was a famous advocate of this diet, which involved living off a liquid protein diet. She lost a lot of weight but later said her championing of the liquid diet was one of the biggest mistakes she'd made on TV.[28] In

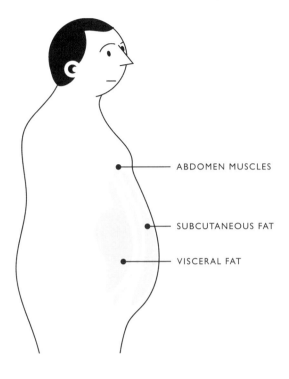

ABDOMEN MUSCLES

SUBCUTANEOUS FAT

VISCERAL FAT

FAT COMES IN DIFFERENT FLAVOURS. VISCERAL FAT, WHICH LEADS TO YOUR WAISTLINE BULGING, IS AN ESPECIALLY DANGEROUS FORM OF FAT.

1995 the Zone diet was promoted by Jennifer Aniston. This involved eating specific ratios of fat, protein and carbohydrate. In 2002 the macrobiotic diet made its entrance. This was supported by Gwyneth Paltrow and involved ditching meat, dairy, eggs, processed foods and sugar and instead having lots of vegetables, legumes and soy. 2007 brought the raw food diet which, yep – involved eating only raw foodstuffs. By 2012 cold-pressed fruit and vegetable juices were all the rage, which may have done nothing for your weight but at least got you a helping of fruit and vegetables. Then in 2014 came the Paleo diet. Celebrities like Jessica Biel and Uma Thurman supported this diet, which showed people how to eat like a caveman or cavewoman: in other words, only eat foods that were available when our ancestors went hunting and gathering, which basically means no processed foods and a good balance of fruit, vegetables, nuts and some meat. Researchers have recently found that at least some of our cavemen ancestors practised cannibalism,[29] which so far at least is not recommended as part of this particular diet. The Paleo diet joined a number of other trends from the past decade, all claiming to help us in all kinds of ways, and none supported by robust science: eat kale, drink bone broth, only use standing desks, eat quinoa, wear a Fitbit and if you're a woman (or maybe even a man) buy a Yoni egg ...

A diet that has recently received a lot of attention is the Mediterranean diet.[30] This came about because of studies of people living in so-called blue zones around the world. These are zones where people live on average to a ripe old age – often hitting 100. Two of these zones, Acciaroli and Sardinia in Italy, have been extensively studied. There are a number of reasons for people living long in these zones, including mild exercise and strong family and community ties, but the diet is felt to be a central part of it. The diet is rich in vegetables (notably legumes such as peas and beans) up to 90 per cent, but also includes herbs such as rosemary, and olive oil. A key feature of the diet is small portions and having the smallest meal in the late afternoon or early evening. People who live into old age in these zones are rarely overweight or obese, and always eat sparingly. [31] Some blue zones encourage a glass of wine a day.

But now, the bad news. There is no scientifically compelling evidence that any of these diets work, or, if they do work and allow you to lose weight, that you will manage to keep your weight down. And some – for example, the grapefruit diet – are plain nonsense. Many of these diets fall into one of three main types: the low-fat diet, the low-carbohydrate diet and the low-calorie diet. The low-fat diet, as the name suggests, is one in which the amount of fat being consumed is dramatically reduced. This type of diet doesn't necessarily mean fewer calories being consumed, just the type of food. The idea is that if you eat too much fat, it ends up being deposited in your body: 'A minute on the lips, a lifetime on the hips' may well be true. The main cell type that holds fat is called the adipocyte. An especially dangerous place to hold fat is around your middle – this is called visceral fat or 'active fat' because it can actively increase your risk of serious health problems.[32] One reason why measuring your waistline is useful is that it will give a measure of how much visceral fat you're carrying. If you're a woman and your waistline is 35 inches or more, you are at risk of health problems. The number for men is 40 inches or more. Fortunately, visceral fat is responsive to exercise and dieting, especially if you shift to a low-fat diet. One interesting finding is that the stress hormone cortisol increases the amount of visceral fat in your body. This is one reason why chronic stress is a risk factor for obesity.[33] A meta-analysis of 16 trials of a low-fat diet running for 2–12 months gave rise to an average weight loss of 3.2 kg and a reasonable proportion of this will be in visceral fat.[34] If you are on this kind of low-fat diet you may have to move your belt out a notch as you will notice a slight shrinking of your waistline.

The low-carbohydrate diet involves cutting down on carbohydrates and eating more protein and fat. You can choose to cut out such sources of carbohydrates as grains, fruit, vegetables and a range of processed foodstuffs that have added carbohydrate, which includes white bread, pasta, biscuits, cakes and sweetened drinks. This diet is thought to work in two ways. The first relies on an interesting piece of biochemistry whereby your body can turn carbohydrates into fat; it's a way of storing nutrients for a rainy day. So if you cut down on carbs, you'll store less fat. The second mechanism involves insulin. After a meal, in response to carbohydrates in your blood, your pancreas releases insulin. This hormone helps your cells to take up carbohydrates to use as a source of energy. When you are hungry, insulin levels are low, and you burn stored fat for energy. A low-carbohydrate diet will lower insulin and, hey presto, you'll burn more fat. Most people can lose weight if they decrease carbohydrate intake. To lose 0.5–0.7 kg a week, you need to reduce intake by 500–750 calories per day,[35] the equivalent of five cans of a soft drink, ten potatoes or ten slices of bread.

A 2015 review found that a diet higher in protein and lower in carbohydrate does seem to help you lose weight somewhat.[36] An interesting extra mechanism is that the more protein you have in your diet, the sooner you will feel full.[37] It's all down to receptors, called mu-opioid receptors, in the walls of the main blood vessel that drains blood from your gut. When they are active, they send a signal to your brain to eat more food. Proteins (or more precisely peptides, which are fragments derived from proteins) suppress the activity of these receptors and so lower your appetite and make you feel full. The low-carbohydrate diet will have various protein sources, including meat, fish and dairy products but also plant-based proteins that occur, for example, in beans and legumes. But some studies have shown that a high-protein diet over an extended period can cause problems such as damage to kidneys.[38] However, most studies indicate that a low-carb high-protein diet can be beneficial for weight loss, at least in the short term.

One popular low-carb high-protein diet is the Atkins diet, named after American cardiologist Dr Robert Atkins, who invented it. The basis of the Atkins diet is as above: to switch from burning carbohydrate to burning stored fat. This switch is known as ketosis, which can be problematic as it can cause headache, fatigue, constipation and bad breath. Evidence indicates that most people who follow the Atkins diet will indeed lose weight but find it difficult to stick with it.[39]

Finally, there is the bog-standard low-calorie diet. This just means cutting down on the total number of calories being consumed and keeping the diet in balance. Again, the recommended cut is 500–1000 calories per day. This can result in a 0.5 kg to 1 kg weight loss per week.[40] The US-based NIC assessed 34 randomised control trials of the low-calorie diet and found that they can reduce total body mass by 8 per cent over 3–12 months, which is substantial.[41] Think about your waistline being 8 per cent smaller – this would mean dropping a size or two in your jeans. There is also an extremely low-calorie diet, which provides 200–800 calories per day and which, like the Atkins diet, is mainly derived from protein and fat. This effectively puts the body into a starvation state and the average person will lose 1.5 kg–2.5 kg per week. A version of this diet, known as 2–4–6–8, follows a four-day cycle with 200 calories consumed on day one, 400 calories on day two, 600 calories on day three and 800 calories on day four.[42] On day five the person fasts for the whole day, and the cycle then repeats. This will result in major weight loss but should only be used for the management of obesity and only under the supervision of a dietician and although there is evidence that some of these approaches work, the rate of slippage once the diet ends is high.

One thing that seems to help is support, and the best example of this support is Weight Watchers.[43] This global company, headquartered in the US, offers a range of products and services to help people lose weight. It is a remarkably successful business, with revenue of $1.5 billion in 2018. (In the same year it rebranded itself as 'WW' to reflect its broader aims of focusing on health and well-being.) The original Weight Watchers was conceived by Jean Nidetch, who had been overweight for most of her life and had tried all kinds of things: pills, hypnosis and lots of fad diets, all of which failed. She joined a weight-loss programme in New York but realised a support network was needed, which initially involved six of her friends, who were all overweight. A weekly weigh-in system was used, and emphasis placed on empathy, rapport, mutual understanding and sharing of stories and ideas.

WW has outlasted all the fad diets. It works by translating calories in food into a points system, which encourages participants to choose nutritious foods, control portions and aim for fewer calories. Many professionals advocate WW. It has the advantage of no junk science or the promise of magic cures. There is a focus on what you should eat rather than what you shouldn't. The social aspect can help people to persist. The big question is: does it work? In 2015 a major study was carried out by Johns Hopkins Medical School.[44] The researchers reviewed 4,200 studies. Of 32 commercial weight loss programmes, 11 had been rigorously studied in randomised control trials – the gold standard to test whether a treatment is working or not. Few studies were scientifically rigorous enough to be reliable, yet WW and another company that helps people lose weight, Jenny Craig, both came out well from the analysis. They both carried out well-controlled trials over 12 months, which provided evidence that the participants achieved greater weight loss than non-participants. Programmes involving the Atkins diet were also supported by trials.

But we are left scratching our heads when it comes to what we should do about the obesity epidemic. Few diets have been scientifically proven to work in a sustained way. The advice from the Johns Hopkins study is that doctors could consider referring overweight patients to Weight Watchers or Jenny Craig, but in reality, those approaches have at best 2–5 per cent greater weight loss than control groups, although this is seen as helpful. The advice that we should all eat less and exercise more is not easy to follow. The incidence of obesity continues to grow, in a manner akin to the way the planet is warming up. It represents the biggest global health challenge of our age in terms of all the diseases it increases the risk of as well as lowering life span. It puts COVID-19 in the shade. Projections are stark: by 2030 nearly one in two adults in the US will be obese with nearly one in four being severely obese.[45] Might the answer come from a better understanding of the genetics underlying the risk of obesity? From that knowledge, will new drugs come along that will truly help us, since we can't seem to help ourselves (and I don't mean to another piece of pie)? Let's hope so, otherwise our future might be an overheated planet inhabited by a population that is either overweight or starving, both groups dying at a faster rate than they otherwise should. **But the bottom line is unrelentingly clear: the vast majority of diets don't work in the long term, so try to eat less and exercise more.**

THE 'BEFORE' AND 'AFTER' PICTURES OF JEAN NIDETCH
(1923–2015). SHE CREATED A SUPPORT-BASED DIETING
PROGRAMME THAT BECAME WEIGHT WATCHERS.

CHAPTER 5

WHY DON'T YOU JUST CHEER UP?

—

'Nothing is funnier than unhappiness,
I grant you that. Yes, yes, it's the most
comical thing in the world.'

—

Samuel Beckett, *Endgame*

HAD A MILD bout of depression when I was 33. My first son, Stevie, had just been born and I had a health scare (which thankfully didn't turn out to be serious), and I think it was a combination of these two things that got me over-ruminating, worrying about not being there for my son and generally feeling hopeless about the future. My sleep was badly disrupted, I had no appetite, and the joy had gone from my life. I would hold my little boy in my arms and think, Why can't I feel the joy of this? Why am I crying? I was lucky though. It was never so bad that I couldn't go to work, but it shook me. I went to see my GP who gave me antidepressants and something to help me sleep. The medication and a bit of talk therapy helped, and I began to come out of it after four months or so, to my relief. I can remember the actual moment when the fog began to clear. We went on a short holiday to Malaga in the spring to escape the damp of an Irish winter. I remember sitting on the balcony of the apartment we had rented, drinking a cup of tea with my mother-in-law, Desiree, in the morning sun. I saw the sunlight sparkle on the sea, and it made me feel good. The episode gave me insight into how others suffer with depression. What happened to me, and why is it that depression has become such a problem for so many of us?

We live in an age when, on the face of it, things have never been better. Men no longer live lives of quiet desperation, spending every waking hour toiling in the fields or down coal mines, dying in wars. Women, at least in western countries, can now live almost as full a life as they want to, free from a cycle of pregnancy and birth, infant mortality and discrimination. And yet the number one fear of teenagers in a recent survey is not the old fear of teen pregnancy or being caught smoking and drinking, but fear of anxiety and depression.[1] A large proportion of adults – as many as 18 per cent – have had at least one major episode of serious depression (defined as requiring treatment – counselling and/or medication) in their lives.[2] And for college students, the numbers seeking help for mental health issues are growing. In Ireland almost 12,000 students sought counselling in 2018, up from 6,000 in 2010.[3] In the US the rate of moderate to severe depression among students rose from 23.2 per cent in 2007 to a whopping 41.1 per cent in 2018. Studies are reporting a prevalence of depression or anxiety at above 35 per cent.[4] College counselling services are overwhelmed. Depression is a serious problem that deserves our closest attention, given the amount of suffering it causes as well as the total tragedy of suicide. Where could it have all gone so horribly wrong? All the promise of good health, enlightenment and freedom has come to a sorry pass. A large proportion of our species are yet again living lives of desperation, although not so quietly as our ancestors.

If you have been depressed, you're in good company. You're in a club that includes Caroline Ahern, Buzz Aldrin, Hans Christian Andersen, Marlon Brando, Kate Bush, Johnny Cash, Leonard Cohen, Charles Darwin, Charles Dickens, Bob Dylan, Stephen Fry, Lady Gaga, Martin Luther King, Stephen King, John Lennon, Abe Lincoln, Spike Milligan, Jim Morrison, Morrissey, Dolores O'Riordan, Sting, Wolfgang Amadeus Mozart, Isaac Newton, Brad Pitt, Sylvia Plath, Edgar Allan Poe, Jackson Pollock, Sergei Rachmaninoff and Bruce Springsteen. And those are just my heroes. They've all had bouts of clinical depression. It's perfectly understandable why Sting would get depressed – anyone who wrote 'Fields of Gold' would have to consider medication. But Brad Pitt? Or Hans Christian Andersen, the writer of those wonderful children's stories that have brought so much joy to the rest of us? This tells us one thing: depression spares no one. And guess what? Success

SPIKE MILLIGAN (1918–2002). COMIC GENIUS, WRITER OF THE 1950S RADIO COMEDY-SHOW *THE GOON SHOW* AND MANY BOOKS. SPIKE SUFFERED FROM BIPOLAR DISORDER HIS WHOLE LIFE, AND YET MANAGED TO BRING SUCH JOY TO MILLIONS IN HIS WRITING AND ACTING.

can bring it on or make it worse.[5] Becoming chief executive officer in a company can induce depression. In fact, CEOs have double the rate of depression when compared to the general public.[6]

So what exactly is depression (also known as major depressive disorder)? It's defined as a persistent state of low mood and aversion to activity. These two features go together. A depressed person will feel low and will also not want to do the activities that they normally do. If someone goes to their GP and says they are suffering from it, the GP will ask them a range of questions. Key indicators are used, the main one being that the person must be suffering from five of the following eight symptoms, and at least one of the first two, nearly daily for at least two weeks: sadness or depressed mood most of the day or almost every day; loss of enjoyment in things that were once pleasurable; a major change in weight or appetite; insomnia or excessive sleep almost every day; physical restlessness that is noticeable by others; fatigue or loss of energy almost every day; feelings of hopelessness or worthlessness, or excessive guilt almost every day; problems with concentration or making decisions almost every day; recurring thoughts of death or suicide.[7] The length of time is what is important here, as we all have these feelings from time to time, but they go away. They will come back, but there will be lots of time without them. Other measures can be used, including the Beck Depression Inventory-II and the nine-item depression scale, the PHQ-9. These measure the severity of the depressed state. Yet no physical or blood test or scan can diagnose depression. Some things can be ruled out by measuring thyroid hormones, as a deficiency can cause depression, but no other test will measure, say, a biochemical in the bloodstream that causes depression.

This makes clinical trials for antidepressants difficult, as whether a drug is working or not can only be evaluated by the patient. Patient reporting (which means a patient filling in a form) is a notoriously uncertain business. Sometimes, patients fill in a form incorrectly or exaggerate or play down how they're feeling. And all the drug tester has to go on is the patient reporting on how they feel. Patient reporting might be one reason why trials for antidepressants have only shown marginal effects. A recent analysis indicates that

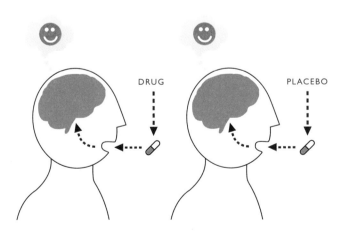

DRUG PLACEBO

PROPERLY RUN CLINICAL TRIALS MUST BE DOUBLE BLIND (MEANING NEITHER PATIENT NOR SCIENTIST KNOWS WHO IS BEING TREATED WITH THE NEW MEDICINE) BUT ALSO PLACEBO-CONTROLLED (MEANING AN UNTREATED PATIENT MUST RECEIVE A 'DUMMY PILL' THAT IS IDENTICAL IN LOOK AND TASTE TO THE MEDICINE BEING TESTED).

most (if not all) of the benefits of antidepressants are due to the placebo response.[8] What is going on here? It seems that just talking to the patient and giving them attention, or the very fact of them being in the trial, lifts their mood. Whatever the chemical imbalance in their brain might be (and the widespread belief is that depression is a chemical imbalance), it might be restored to normal by medication, or by a range of other non-medical interventions, such as a placebo. It's important to stress that placebo is not an absence of intervention. It's an intervention because the patient knows they are in a trial, even though they aren't being treated directly.

Major depressive disorder is a common illness. The percentage of sufferers varies from country to country, with an occurrence ranging from 7 per cent in Japan to 21 per cent in France.[9] In most countries the incidence ranges from 8 to 18 per cent, depending on the study. This means that in a room of 100 people, somewhere between 8 and 18 will have suffered at some point in their lives from depression. Major depression is twice as common in women as it is in men, although it is not clear why this is the case. People are most likely to develop their first episode of depression between the ages of 30 and 40, with another smaller peak between the ages of 50 and 60. The risk of depression significantly increases if you have an underlying neurological disorder such as Parkinson's disease, stroke or multiple sclerosis, following a heart attack, and during the first year after childbirth. Postnatal depression is a serious disorder affecting 10 to 15 per cent of women.[10] It is thought to be caused by the hormonal changes that happen during and after childbirth and the pressures of being a parent. Several studies have shown that the incidence of depression is lower in the elderly, although the reason for this is not known. One theory is that as you get older, you gain more perspective on life and you no longer sweat the small stuff.

The centre of depression is in our brain, although the Ancient Greeks thought it was to do with the bile duct. They thought that our mental state was driven by the balance of four fluids or 'humours' in our bodies: black bile, yellow bile, phlegm and blood. Black bile was the one you didn't want out of

Sanguine	Choleric	Melancholic	Phlegmatic
Blood	*Yellow bile*	*Black bile*	*Phlegm*

kilter. The Greek word for black is *melan* – this gave rise to the term 'melancholia', which referred to the idea that an imbalance in black bile was the cause of depression. The term 'humour' (as in mood) also comes from this idea. We now know it is the mind that counts in depression, and the mind is located in the brain. That being said, we still don't know what the mind is. At its simplest, it's to do with neurons in our brains forming complex circuits, although neuroscientists warn us not to think of it as a computer. It's much more complicated.

THE ANCIENT GREEKS BELIEVED THAT MOOD WAS CONTROLLED BY THE BALANCE OF FLUIDS IN OUR BODIES. THIS LED TO THE THEORY OF THE FOUR TEMPERAMENTS: SANGUINE (OPTIMISTIC), CHOLERIC (PASSIONATE), MELANCHOLIC (SENSITIVE) AND PHLEGMATIC (COMPOSED).

There are 100 billion neurons in your brain, all cracking and firing. It is the interplay between these neurons that is thought to explain things like memory, intelligence and personality, although we're clueless as to how this complex cellular biochemical machine might work. The assumption when it comes to depression is that some imbalance happens in the workings of our minds, which leads to the changes that characterise this disease. Results from attempts to measure differences in the brains of depressed people have revealed some differences. Depressed patients have been shown to have increased volume in a part of the brain called the lateral ventricles and smaller volumes in other brain regions, such as in the thalamus, hippocampus and frontal lobe.[11] This suggests that there are indeed changes in brain structures in depressed patients. Scans that measure brain activity also show some differences in depressed patients, but results are mixed, and brain scans are not routinely used to diagnose depression or follow its course.

Neurons connect to each other by releasing chemicals called neurotransmitters. These are like batons in a relay race with one runner (the neuron) passing the baton (the neurotransmitter) to the next runner. There are lots of different types of batons, including serotonin, noradrenaline, acetylcholine, dopamine and glutamate. And although we talk about some of these, such as serotonin, as being disturbed and needing to be restored with drugs, evidence that they are off-kilter or that they actually change in someone taking antidepressants is largely missing. In one way, this is outrageous. Ed Bullmore, professor of psychiatry at the University of Cambridge, has written that when a depressed patient asks a psychiatrist what is wrong with them, the psychiatrist will tell them about disturbed brain chemicals and how the drugs can fix them. But when the patient presses the doctor, asking if these chemicals can be measured, either to help with diagnosis or to show that a treatment is working, the doctor will say no.[12] Both doctor and patient take these things on faith to some extent – hardly ideal for the twenty-first century, where we think science is all-pervasive. There is evidence, of course, for neurotransmitters affecting mood, but this has been shown mainly in animals, or based on genetics, where genetic variants that give rise to proteins that regulate levels of neurochemicals, such as serotonin, can be linked to depression. Nothing definitive has emerged for doctors to use diagnostically to confirm that a patient is depressed and to indicate a course of action to help that patient.

Serotonin is a fascinating neurotransmitter. It is of interest here because serotonin selective reuptake inhibitors (SSRIs) are a mainstay treatment for depression. SSRIs include drugs like Prozac, the wonder drug of the 1990s. In 2017, almost 22 million prescriptions for Prozac were written in the US, with a similar number over the previous 10 years.[13] A recent study revealed that the number of prescriptions of antidepressants in England has almost

doubled in the past decade. In 2018, 70.9 million prescriptions were given out compared to 36 million in 2008.[14] SSRIs are the main type prescribed. Other antidepressants include monoamine oxidase inhibitors. These block the breakdown of monoamines, including serotonin, in the brain. In Ireland, prescriptions are up by two-thirds since 2009, with Ireland's most common antidepressant, Lexapro (another SSRI),

INSIDE OUT (2015) IS A FILM ABOUT THE MIND OF A YOUNG GIRL. HER EMOTIONS ARE CONTROLLED BY 5 CHARACTERS: JOY, SADNESS, ANGER, FEAR AND DISGUST.

being prescribed 609,655 times in 2017.[15] That's an awful lot of tablets for an awful lot of depressed people. But do the drugs work? They certainly seem to work for some, with an overall response rate of a 50 per cent reduction in depression scores for moderate to severe depression.

Serotonin affects approximately 40 billion brain cells: this includes brain cells involved in mood, sexual desire, appetite, sleep, memory and learning, social behaviour and even temperature regulation in your body. There are, however, no ways to measure it in a living brain, although this may be because any changes might only be evident in a tiny part of the brain. And there is no evidence linking the level of serotonin to depression or any mental illness. It can be measured in the blood, and there is some evidence of it being lower in the blood of depressed people, but that might be due to the depressed state rather than the drop causing depression. The bottom line is that SSRIs supposedly work by increasing serotonin levels in the brain, but exactly how they work is not fully understood. Another one of the big mysteries of SSRIs is why they can take a month or more to have an effect. Nor is there evidence that exercise – a proven way to treat depression – boosts serotonin levels. The FDA carried out a systematic review of clinical trials, which revealed that antidepressants stave off depression by 52 per cent overall,[16] and a major study revealed that in 522 trials anti-depressants were more efficacious than placebo.[17]

Currently, brain chemistry and functioning let us down when it comes to a physical explanation for depression. Another key area where there are ideas but no clear conclusions is the question of what causes depression in the first place. Again, the comparison with other diseases is stark. We have a much better idea what causes, say, infectious diseases, which are caused by bacteria or viruses, or Type 1 diabetes, which is caused by lack of insulin, or cancer, which is caused by mutations in genes that control the growth of cells or that prevent tumours from growing. Treatment can then be deployed based on this knowledge. But for depression, the causes can be many and varied.[18] Clearly, life events play a major role. These include bereavement, financial difficulties, unemployment, a medical diagnosis, bullying, rape, social isolation, romantic disasters, major injury or even success, as mentioned above. These events trigger a sense of loss – be it the loss of a loved one, or our health, or our freedom. Or they make us worry, which gives

rise to rumination, which then leads to a depressive episode. Can it be that all these things lead to an imbalance of neurochemicals in our brains that can then be fixed with a drug? This seems unlikely, and yet that is the best we've got as an explanation. For many, the depression eases with time, or we learn to live with the feelings we have – our brain adapts to the new circumstances. But many people need treatment and, most important, should be encouraged to get help. Even though we might not know the exact cause, and even though we might not be able to measure things in the brain or body of someone who is suffering, everyone must be encouraged to seek help, since depression, like any other disease, can be helped.

Although many different life events can cause depression, adversity in childhood, particularly physical or sexual abuse, is a major predictor of depression later in life. Drugs, including alcohol, sedatives and stimulants such as cocaine or amphetamines, can also either exacerbate or cause depression. These drugs affect the brain, and during withdrawal, it is thought that the brain compensates for being disturbed, and this compensation can somehow lead to depression. Perhaps the drug lowers neurotransmitters linked to anxiety,

**THE CHEMISTRY BEHIND
YOUR HANGOVER ANXIETY**

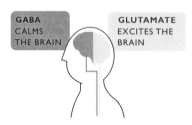

GABA
CALMS
THE BRAIN

GLUTAMATE
EXCITES THE
BRAIN

WHEN WE DRINK...

GABA INCREASES
AND GLUTAMATE
IS BLOCKED

THE NEXT MORNING...

THE BRAIN TRIES
TO CORRECT THE
IMBALANCE

THE GLUTAMATE REBOUND

THE GLUTAMATE BOUNCES
BACK TO A HIGHER LEVEL,
GIVING RISE TO ANXIETY.

ALCOHOL IS THOUGHT TO BOOST THE
CALMING NEUROTRANSMITTER GABA
WHILE SUPPRESSING THE EXCITATORY
NEUROTRANSMITTER GLUTAMATE. THIS
IMPROVES YOUR MOOD. AFTER A NIGHT ON
THE LASH, HOWEVER, THE BRAIN FIGHTS
BACK. GLUTAMATE GOES BACK UP, BUT
OVERSHOOTS. RESULT: HANGXIETY, ONE
OF THE SYMPTOMS OF A HANGOVER, THE
MEDICAL TERM FOR WHICH IS VEISALGIA
(WHICH MIGHT COME IN USEFUL IF YOU
WANT TO GIVE YOUR BOSS THE REASON
WHY YOU CAN'T COME TO WORK).

and during withdrawal from the drug, these neurotransmitters bounce back but overshoot, causing the anxiety. During a hangover, when the alcohol wears off, the euphoria is over and the parts of your brain that were switched off by the alcohol (which is thought to be those parts of the brain that provoke feelings of anxiety) become more active as they rebound, so you feel way more anxious and depressed until everything returns to normal. These events even have a name – they are called 'the glutamate rebound'.[19] Glutamate is an excitatory neurotransmitter, which means it stimulates the brain, and alcohol is thought to lower the level of it (which is one reason why anxiety eases when we drink), but glutamate then bounces back to a higher level, giving rise to anxiety. There is even a term for the anxiety that comes with a hangover: 'hangxiety'. People who are shyer and more introverted have been shown to have much higher levels of anxiety during a hangover than the people who weren't shy to begin with. Overall, alcohol can have a profound effect on people prone to depression. Alcoholics are also prone to depression. Life events can cause people to become alcoholics, and then the alcohol leads to an exacerbation in the underlying depression.

The genetic basis for depression has also been an area of intense investigation.

Family and twin studies have revealed that almost 40 per cent of the risk of becoming depressed is down to our genes.[20] This is a high percentage when it comes to implicating genetics in a given condition and indicates that if a near relative has had depression, there's a reasonable chance that you will too. The overall risk of depression in the general population is about one in four, but if your parents have been depressed that risk jumps threefold, with grandchildren of depressed grandparents also being at an increased risk.[21] Children of depressed parents have a 75 per cent chance of being depressed. Having a sibling with depression increases your chance two- or threefold. Identical twins, who share 100 per cent of their DNA, are at a higher risk of developing depression than fraternal twins, who only share 50 per cent. This is important as the reasonable assumption is that both sets of twins will be raised in the same environment. If it were purely environmental, then both identical and fraternal twins would have the same risk of depression, which is not the case. However, depressed parents or siblings might create an environment where depression is more likely, and so the environment will also have an influence. Depression is thus likely to be the result of a complex interplay between genes and environment.

The genetics behind depression is complex and it's difficult to pinpoint which genes are involved. In 2019, 102 variants in genes that increase the risk of depression were identified.[22] Genes like 5-HTTLPR, CRHR1 and BDNF have variants that have been linked to the risk of depression, which means if you carry those particular variants, your risk is higher than someone with a different variant. Given their function, the link is somewhat plausible: 5-HTTLPR is known to control serotonin (which SSRIs boost), CRHR1 has been linked to how the body responds to stress (a cause of depression) and BDNF helps neurons grow. One theory of depression posits that an impairment in neuronal growth might be a cause. Neurons die in our brains all the time, and recent studies suggest that they are replaced. A defect in this replacement might be a reason for depression but these genetic associations have proven difficult to confirm in subsequent studies, and so the jury is still out.

Another study may be more robust: in this major study involving multiple centres around the world, out of 20,000 genes, 44 have been identified that increase the risk of depression.[23] This number is important but equally presents a challenge, because it is highly unlikely that there will be a single gene variant. Depression will involve multiple genes, each contributing a small amount to the overall risk. Several of the genes code for proteins involved in brain function, and so it's no surprise that they might be involved in depression; this provides yet more evidence that the roots of depression lie in the mind. It is likely that one day it will be a combination of genetic variants that will be used to predict the risk of depression and even point to possible treatments. It will, however, most likely be a combination of carrying certain variants of genes in a particular environment or in someone who has suffered in various ways that will lead to depression.

Why college students are at a higher risk of depression has also been the focus of much analysis. Risk factors are thought to include pressures linked to social media, personal and family expectations over academic success, and poor sleep quality. One study has found that almost 50 per cent of college students indicated that they woke up during the night to answer text messages.[24]

Treatments for depression vary widely. In the UK, the recommendation is that antidepressants, with SSRIs being the main option, should not be used initially, especially in the case of mild depression, because the risk–benefit ratio is poor (risk in this case meaning side effects). SSRIs should be used for moderate or severe depression and should be continued for at least six months. Apart from medication, doctors can recommend psycho-therapy, exercise (which has been shown to be as effective as medication or talk therapy) and when the depression is severe and not responding to other approaches, electroconvulsive therapy (ECT). This involves electro-cuting the brain and has been shown to bring remarkable results in some patients, with a response rate of 50 per cent in patients who don't respond to other treatments. It is especially useful in the severely depressed, but exactly how it works is unclear. Cognitive behavioural therapy (CBT) has the most evidence for being efficacious.[25] It consists of teaching patients to challenge their thinking patterns and also change counterproductive behaviour, and it appears to be especially effective at preventing relapses. The programmes that best prevent depression need more than eight sessions, each lasting between 60 and 90 minutes. A Dutch programme, known as the 'Coping with Depression' course, has seen notable success, reducing risk by 38 per cent.[26]

Pioneered by Sigmund Freud, psychoanalysis is an approach that tries to resolve unconscious mental conflicts, which are revealed upon questioning by the therapist. It must be stated that there is no evidence for any of the ideas that Freud came up with about the mind. We can't detect the unconscious mind or confirm scientifically any of his theories about the mind, in which he famously came up with the idea of the id and the ego. (Freud is also alleged to have said that the Irish are the only people who are impervious to psychoanalysis.) There is evidence that psychotherapy is beneficial, at least as beneficial as medication for mild to moderate depression. Psychotherapy and cognitive behavioural therapy may simply be another example of the placebo effect in action.

On average, a depressive episode will last three months, so there is always a light at the end of the tunnel.[27] Yet there is a risk of recurrence, with 80 per cent of people having at least one more episode in their lifetime, with a lifetime average of four episodes. Around 15 per cent of people suffer chronic recurrence. The recommendation is for patients to continue on antidepressants for four to six months after recovery, as this will reduce the chance of relapse by 70 per cent. Sadly, depression is a risk factor for suicide. Up to 60 per cent of people who die by suicide have had a mood disorder but the overall risk of a depressed person committing suicide is quite low, standing at a 2 per cent risk for those ever treated in an outpatient setting. This rises

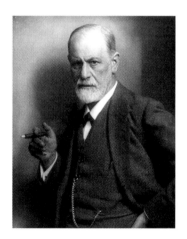

SIGMUND FREUD (1856–1939). THE DADDY OF THEM ALL WHEN IT COMES TO PSYCHOANALYSTS. NO SCIENTIFIC BASIS FOR HIS THEORIES HAS BEEN FOUND.

to 4 per cent for those treated as an inpatient, indicating that the risk is related to severity of depression. There are also gender differences, with about 7 per cent of men with a lifetime history of depression dying by suicide, compared to 1 per cent for women (although suicide attempts are more common in women).[28] The suicide rate in the general population stands at about 0.01 per cent.[29] Interestingly, the rate of suicide is dropping in most countries, with notable decreases in Russia where between 2000 and 2012 it fell by 44 per cent, in the UK by 21 per cent and in Ireland by 38 per cent.[30] One country bucks this trend: in the US it increased by 24 per cent in that period, and it's not fully known why.

An important question is whether the incidence of depression is on the rise. Several studies are underway into the mental state of so-called millennials – people born in the 1980s and 1990s. Studies indicate that this generation are much better educated than their parents, less hedonistic, better behaved – but also lonelier than previous generations; and this in a world that has never been more connected, what with smartphones and social media. The eminent Pew Research Center carried out a poll of 920 Americans aged 13–17 about problems among their peers.[31] They are far less concerned about things like their parents finding out that they drink alcohol, or unwanted pregnancy. Instead, 70 per cent of respondents said anxiety and depression were the most important issues among their peers. Fifty per cent also said that fear of drug addiction was a major concern. Issues that are worrying teens include fear of letting down their parents (since parents are more connected than ever to their children) and also the pressures of social media.

Where are we likely to see progress when it comes to treating depression in the future? An interesting development concerns the emerging role of the immune system as a cause of depression. At first, this might seem surprising – your immune system is all about fighting infection by using white blood cells that home in on and attack bacteria, viruses and parasites. Why would such a beneficial system make you depressed?

This idea began with the realisation that when you have a cold or a flu

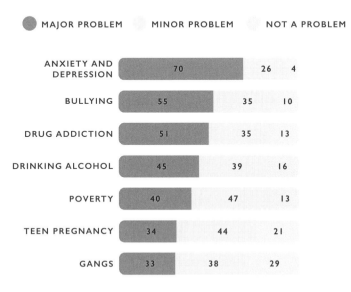

ANXIETY AND DEPRESSION TOP LIST OF
PROBLEMS TEENS SEE AMONG THEIR PEERS

% OF TEENS SAYING EACH OF THE FOLLOWING IS A _____ AMONG
PEOPLE THEIR AGE IN THE COMMUNITY WHERE THEY LIVE

● MAJOR PROBLEM   MINOR PROBLEM   NOT A PROBLEM

	MAJOR PROBLEM	MINOR PROBLEM	NOT A PROBLEM
ANXIETY AND DEPRESSION	70	26	4
BULLYING	55	35	10
DRUG ADDICTION	51	35	13
DRINKING ALCOHOL	45	39	16
POVERTY	40	47	13
TEEN PREGNANCY	34	44	21
GANGS	33	38	29

SURVEY OF US TEENS AGES 13 TO 17 CONDUCTED
17 SEPTEMBER–25 NOVEMBER 2018

PEW RESEARCH CENTER

you are inclined to show the symptoms of depression: these include lack of appetite, social withdrawal (you crawl under the duvet) and a low feeling (only exacerbated by watching daytime TV). It's likely that this response is an evolved one: it allows your body to recuperate but also removes you from active contact with the rest of the herd so that you don't spread the infection to others and wipe out the whole community. This response goes by the name of 'sickness behaviour' and is common in infection but also in inflammatory diseases such as rheumatoid arthritis or Crohn's disease. For unknown reasons these diseases of the immune system attack your own tissues, which in the case of arthritis means your joints and in Crohn's disease means your digestive system. So what is going on here? The sickness symptoms are being driven by immune molecules called cytokines. These are the wake-up call of the immune system, mobilising the troops to fight the invader, but some of them cause sickness behaviour. A good example are cytokines called interferons, molecules that are made in response to viral infections. Interferons are good at promoting the killing of viruses and are sometimes used to treat patients with viral infections, such as hepatitis C. But when doctors gave patients interferons, they noticed that the patients were inclined to become depressed. And then doctors working on rheumatoid arthritis noticed the 'Remicade High'. Remicade is a treatment for rheumatoid arthritis that works by blocking another cytokine, called TNF. This cytokine is important in arthritis as it causes many of the symptoms of that disease, including joint pain and destruction. For some patients, blocking TNF relieves these symptoms. But doctors noticed that patients not only rapidly felt better on the treatment – they also had more energy, their 'brain fog' cleared and their mood lifted. Again, this indicated that TNF was responsible for these symptoms. Doctors used to think that the depression that goes along with diseases like arthritis was because of sore joints and an inability to do things that gave pleasure before, like taking exercise. But it turns out they were wrong: TNF also has effects in the brain and promotes sickness behaviour. The depression that is a feature of infections, which is designed to stop the infection spreading and makes sure you rest, goes rogue in inflammatory diseases, because in both cases cytokines are being made. All of this strongly points to cytokines as a new target to go after in fighting depression.

Could 'regular' depression, which happens in people without infections or inflammatory diseases, also involve cytokines? The trigger here might actually be stress, which is a major risk factor for depression. It turns out that stress causes inflammation in our bodies too, just like an infection or injury. This provides another missing link – stress will elevate certain cytokines that then cause depression. The link between stress and cytokines is likely to be due to evolution, as the stress response evolved in part to respond to danger.

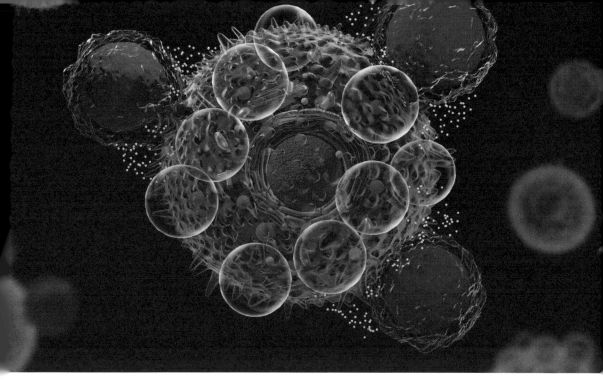

During a moment of danger (e.g. an encounter with a tiger) you might get hurt, so your immune system is at the ready to fight any infection that might come with the injury and then repair you. The problem is, the stressors we now have aren't necessarily tigers, but instead are the threat from your angry boss or argumentative friend or annoyed spouse. That stress is then interpreted as a cytokine response, and that might in turn lead to depression.

CYTOKINES (THE SMALL YELLOW DOTS) ARE MESSENGER MOLECULES MADE BY YOUR IMMUNE SYSTEM THAT HELP YOU FIGHT INFECTION. SOME OF THEM, HOWEVER, CAN CAUSE DEPRESSION. THEY MAY BE A NEW TARGET IN THE EFFORT TO DEVELOP BETTER ANTIDEPRESSANTS.

This idea that depression is some kind of inflammation of the mind is creating great excitement, and clinical trials are underway to test how useful cytokine targeting might be. They might give rise to a whole new class of antidepressants based on strong scientific evidence: such drugs would represent the first major advance in the treatment of depression in 30 years. Apart from the prospect of new therapies, work on the link between the immune system and depression is providing new insights into what is going wrong. Of the 44 genes associated with risk of depression, several are in the immune system, providing further evidence for an immune component and further justifying the targeting of immune factors to treat depression. Immune system proteins might also prove useful in the diagnosis of depression. In one study, patients with major depressive disorder who have high suicidal ideation had higher levels of certain inflammatory proteins than those with low suicidal ideation.[32] These kinds of studies might give rise to tests to identify depressed people who are especially at risk of suicide, which could prove to have major benefits by preventing the tragedy of suicide.

Another area generating great interest is the idea that gut bacteria are a cause of depression. It's been shown that in people with major depressive disorder, the bacteria in their gut differs from that in people without depression, with several species missing. When the mix of bacteria that are there are transplanted into rats, the rats become depressed.[33] The missing bacteria might be a new treatment for depression.

Two other new approaches are also of interest and come from another most unexpected place: recreational drugs. The drug ketamine is a powerful tranquiliser, used clinically on farm animals requiring surgery. Its secondary use, at lower doses, is as a party drug, creating what is known as 'a dissociative state' in the minds of those who take it. But evidence also indicates that it could relieve the symptoms of depression and clinical trials confirm this.[34] What makes it different to SSRIs is that it acts rapidly – patients don't have to wait a month to see any benefit. How it works is not known but it has been called the single most important advance in treating depression in 50 years.[35] Ketamine was recently approved as an antidepressant by the FDA for patients not responding to other antidepressants. Let's see how useful it will be.

Psilocybin is another drug which might have anti-depressant properties. Guess what this is the active ingredient in? Magic mushrooms. These give people who use them psychedelic experiences, similar to LSD. However, low-dose psilocybin has also been shown to relieve depression and it is currently being extensively studied.[36] It may provide doctors with another weapon to use in their efforts to come up with a treatment for patients who are suffering. People with depression shouldn't take magic mushrooms outside of a clinical setting as the amount of psilocybin will vary from mushroom to mushroom, and they may end up exacerbating their condition. But psilocybin could provide doctors with another weapon to use in their efforts to come up with a treatment for patients who are suffering.

Given the increasing incidence of depression, especially among the young, and given how it robs people of the joy of life, the disease needs our closest attention. We've only got one life, and we must do everything we can to live it to the full and help those who are struggling. There are, in fact, a range of things that lift people's moods and help stave off the dreaded black dog, as Churchill used to call it. We're told that in order to be happy we actually need to make an effort. This could be because our brains are inclined to default into worry and anxiety as a protective mechanism against harm. We're also inclined to overthink things, again presumably as a survival mechanism.

But we are not powerless. We can fight these tendencies. Some are obvious, such as trying to maintain a positive outlook, hanging out with positive people and being part of a positive community. All of these help us not to descend into a slough of despond. Equally, regular exercise and enjoying the natural world have all been shown to bring benefits. In one enormous study involving 1.2 million people, on average people reported 3.4 days per month of poor mental health.[37] This was reduced by 1.5 days in those who took regular exercise. Having a pet has also been shown to be remarkably beneficial in improving our mood. Pets give us a focus, get us to take exercise and are an opportunity to socialise with others. Interacting with your pet dog has been shown to boost oxytocin, the love hormone, in both the dog and the owner. Volunteering has also been shown to be remarkably effective, as has having a job that gives us meaning (which usually means a job that helps our fellow human beings). Ultimately, there are things we can do either to stave off depression or to stop it happening again. **The bottom line: by following the guidelines above and with help from your doctor, you can beat depression. You're not here to live in pain and fear. Depression, be gone!**

CHAPTER 6

WHY CAN'T YOU STOP DOING THINGS THAT ARE BAD FOR YOU?

—

'There are all kinds of addicts, I guess.
We all have pain. And we all look for ways
to make the pain go away.'

—

Sherman Alexie, *The Absolutely True
Diary of a Part-Time Indian*

ET READY FOR a shock-horror admission. I have taken drugs. Caffeine. Alcohol. And yes, cannabis. I tried magic mushrooms, too, and came out the other end loving Pink Floyd. Biggest mistake of my life. We humans are a curious species, and we like trying new things. We also keep developing things we become addicted to and that cause us trouble. The latest addiction is your smartphone. I recently lost mine, and it felt like I'd lost a limb. I felt intensely uneasy and couldn't stop thinking about the loss. I did everything I could to get a replacement. I felt great relief when I bought a new one and uploaded information from the Cloud (back up your iPhone right now, fellow addicts). A study into 'problematic smartphone usage' among teenagers has concluded that as many as 30 per cent have a problem that looks remarkably like an addiction. Research commissioned in 2019 by Deloitte showed that Irish people on average check their smartphones 50 times a day – higher than the European average of 41.[1] Smartphones have been shown to light up the reward centres in your brain, giving you a dopamine hit every time you get a notification, not unlike how addictive drugs work. Like other addictions, smartphones are all about seeking pleasure, and the list of substances and activities that do this continues to grow. Addiction is a scourge on people's lives and every government worries about it. It has an astonishingly high financial and emotional toll. Why are we built this way, and what can we do when something becomes so addictive that it ruins our lives?

Although smartphone addiction is nowhere near as serious as an addiction to heroin or alcohol, studies on smartphone use are telling us a lot about the nature of addiction. For most of us, the alarms on our smartphones wake us up – and then the stream of notifications begins (as we saw in the last chapter). One possible reason for the upsurge in anxiety and depression in young people is that one-third of teens wake up in the middle of the night to check their smartphone.[2] This is highly disruptive to their sleep, a well-known cause of anxiety. The apps and social media sites we use feed our need for social contact, information and fun. We've become hardwired to our phone. Here are the features of problematic smartphone usage, which are strikingly similar to how addiction is defined – have a look at these and see if you have any:

- You regularly have an intense urge to use your smartphone.
- You spend more time on it than you intend.
- You panic when the battery runs down.
- You keep using it even though you know it's having a negative effect on your life (that's a big one when it comes to addiction).

Studies show that 89 per cent of university students report feeling phone vibrations that aren't there.[3] As many as 86 per cent of us regularly check our emails and social media sites, on average 55 times a day.[4] All of this suggests addiction, and recently brain scanning would appear to confirm this. Scientists have used a technique called magnetic resonance imaging spectroscopy to examine the brains of teenagers who were diagnosed with internet or smartphone addiction.[5] Although the numbers were small – 19 were in the 'addicted' group and 19 were in the control group – interesting findings were made. The control group in this case comprised teenagers who were matched for gender and age but were not addicted. (How they found the control group in this day and age remains a mystery.)

A questionnaire was used to assess the extent to which internet or smartphone use affected daily lives, social activities, productivity and sleep

patterns. The addicted teens had higher scores for depression, anxiety and insomnia. What the scientists found was fascinating. They were able to measure certain brain chemicals – GABA and glutamate – and observed major changes in a part of the brain known as the anterior cingulate cortex, which confirm the profound effect smartphone use has on our brains. Importantly, 12 of the addicted group received nine weeks of cognitive behavioural therapy, modified from a programme for gaming addiction. When the teens went through the therapy sessions, brain chemicals returned to normal. This study indicates that smartphone or internet addiction is causing changes to our brain neurochemicals, and such changes lie at the heart of any addiction.

What actually constitutes addiction? It is defined as a brain disorder whereby people engage compulsively with a rewarding stimulus despite adverse consequences. You know it's bad for you, but you keep doing it. This seems stupid, as surely we should be able to stop doing things that are bad for us? And yet it seems that the part of our brain that controls common sense or insight doesn't seem to work in those of us who are addicted, or is overridden by those parts that drive addiction. It's almost as if we have an angel and a devil on our shoulders and the angel is ignored in favour of the devil. The devil of addiction has two features. First, its use is reinforcing, which means that when you partake of the substance or behaviour, you will be driven to seek repeated exposure. Second, what you are doing is perceived as being rewarding – it makes you feel good, at least when you're using the addictive substance or participating in the addictive behaviour. Addiction is different from dependency, which is defined as a disorder where cessation results in an unpleasant state known as withdrawal, which usually involves symptoms such as irritability, fatigue and nausea. These may provoke further use, but withdrawal symptoms are different from the compulsive behaviour of addiction, in which a person unrelentingly seeks the substance or activity to which they are addicted, which often occurs independently of any physical symptoms. Addiction and withdrawal, however, often occur together.

Overall, there are two main categories of addiction – chemical and behavioural.[6] Chemical addiction means that you are addicted to a substance, whereas behavioural addiction means you are addicted to an activity. Chemical addiction is extremely common in Ireland, be it to alcohol or the whole gamut of so-called drugs of abuse. Around 40 per cent of Irish people binge-drink, which is much higher than most other countries.[7] Binge drinking is defined for men as consuming more than six standard drinks (the equivalent of three pints or three medium (175ml) glasses of wine) over a period of two hours. For women, it's four standard drinks in the same period. This level of alcohol consumption is considered bad for us.

Alcohol is poison in our bodies and the liver does a great job at breaking it down. The problem is, our livers can only handle one standard drink per hour, so if you drink more than that it can't cope, and the alcohol starts to cause damage. Our brains are hypersensitive, and one reason why we feel euphoric is the parts of the brain involved in anxiety (especially social anxiety) are turned off by alcohol, so we relax and enjoy ourselves.[8] This is where alcoholism begins. The changes in brain chemistry can set us on the path to addiction. Long-term binge drinking increases the risk of liver disease, given the hammering we are giving our livers, which eventually become so scarred by the injury that they pack up altogether. It also increases the risk of various types of cancer, notably those parts of the body that are most exposed to the alcohol: the liver, mouth, throat, oesophagus and digestive system. There is also a much higher risk of heart disease and stroke. Ireland has the second highest rate of binge drinking in the world, with 81 per cent of the population consuming two and a half times the global average of alcohol.[9] The Irish seem to be world-beaters when it comes to alcohol abuse. Around 150,000 people in Ireland are alcoholics. The reasons for excessive alcohol consumption in Ireland are varied and have been reported to include the malign influence of the Catholic Church, English colonialism and even the weather. Alcohol is a major part of social and cultural life in Ireland, and, as such, it is difficult to avoid.

In spite of these statistics and recommendations, it's not clear what binge drinking actually does to the Irish population. Life expectancy overall is not that different from other countries, nor is the incidence of a range of diseases associated with alcohol consumption. The population that binge-drinks into old age don't seem to be any more infirm than one that doesn't. The fact that so many people consume over two and a half times the average global level of alcohol isn't necessarily translating into two and a half times the rate of alcohol-related diseases. The Irish population are no more likely to die of liver cirrhosis than other European countries, but a particularly worrying trend is the number of women who are seeking help for alcohol addiction; many of these women also have a co-addiction to prescription medications for depression or anxiety.[10]

Looking at other chemicals to which we risk becoming addicted, nicotine is next in overall use.[11] Nicotine, found in the tobacco plant, acts on nicotinic acetylcholine receptors in the brain. The tobacco plant most probably makes it as an insecticide, a separate feature to being an addictive drug in humans. It is classified as a stimulant; it triggers the release of dopamine, giving pleasurable feelings. It also causes the release of another neurotransmitter called epinephrine (also called adrenaline), which gives you a buzz. However, once you are addicted, you need nicotine just to feel OK – the very nature of what it is to be an addict. The Healthy Ireland Survey

THE TOBACCO BRAND 'TOBACCO KILLS' NEVER TOOK OFF.

MARIJUANA IS NOW
LEGAL IN MANY PLACES
AND CAN BE SOLD
IN SHOPS LIKE ANY
OTHER AGRICULTURAL
PRODUCT.

found that around 17 per cent of Irish adults are smokers, with 14 per cent smoking on a daily basis.[12]

Each country in Europe produces an annual report on drug addiction under the auspices of the European Monitoring Centre for Drugs and Drug Addiction (EMCDDA).[13] The report for 2019 revealed that drug use has become more common in Ireland in 15–64-year-olds. Over the past 20 years, drug use in Ireland has been increasing steadily. In 2002 fewer than two in ten adults reported using any illicit drug in their lives, but this number has risen to three in ten in 2014. Between 2016 and 2017 the number of cases of cocaine addiction jumped by 32 per cent. Cocaine works by preventing neurons from taking up the neurotransmitters serotonin, norepinephrine and dopamine, boosting the levels of all of these.[14] It's like turning on all the lights in your brain at once, creating a feeling of euphoria. MDMA (also known as ecstasy) works in a similar way.[15] Cannabis remains the most used illicit drug in Ireland, followed by MDMA and cocaine.[16] Cannabis has the active ingredient tetrahydrocannabinol (THC), which, like nicotine in tobacco, acts as an insecticide for the plant. The insects that feed on cannabis leaves get stoned and fall off. THC binds to cannabinoid receptors in the brain and induces feelings of relaxation and, to a lesser extent, euphoria.[17] Of those who seek treatment for cannabis addiction, 79 per cent are male and 21 per cent are female.

The numbers are similar for cocaine with 81 per cent seeking treatment for cocaine addiction being male and 19 per cent being female. For young adults in the age bracket 15–34, 13.8 per cent used cannabis, 4.4 per cent used MDMA, 2.9 per cent used cocaine and 0.6 per cent used amphetamines. Amphetamine increases alertness, concentration and self-confidence. It boosts the levels of dopamine and norepinephrine in the brain by stopping the enzymes that break them down.[18] In 2017 there were 18,988 heroin users in Ireland, of which 10,316 attended opioid substitution treatment centre. Heroin works by binding mu-opioid receptors, which act to limit the inhibitory neurotransmitter GABA, causing euphoria. Ireland places slightly above the European average for the use of these drugs.

There are also problems with addiction to prescription drugs. Prescription-drug abuse is defined as the use of a medication without a prescription in a way other than as prescribed. Several studies in Ireland have found that medicines that are used to treat pain, attention deficit disorder (defined as poor concentration, hyperactivity and learning difficulties) and anxiety are being abused at a rate second only to cannabis.[19] It is likely that these drugs are being obtained illegally or on the internet. In the US in 2016 prescription drugs were implicated in 70 per cent of poisoning deaths.[20] The opiate methadone (used to help people kick heroin) and the tranquilliser diazepam were the most common prescription drugs implicated, with methadone being implicated in 30 per cent of these cases.

One of the most troubling stories about prescription-drug addiction is the epidemic in the US of addiction to oxycodone (trade name Oxycontin).[21] Like other opiates, such as heroin, it acts on the mu-opioid receptor and is used as a powerful painkiller in cases of moderate to severe pain. It has an intriguing

THEODOR MORELL (1886–1948), HITLER'S DOCTOR. HE KEPT HITLER HIGH ON THE OPIATE OXYCODONE UNTIL SUPPLIES RAN OUT IN 1945, POSSIBLY EXPLAINING HITLER'S BEHAVIOUR TOWARDS THE END OF WORLD WAR II.

history. Oxycodone was first made in Germany in 1916. The German pharmaceutical company Bayer had made heroin in the late 1800s. It was actually used as a cough medicine before Bayer realised the other rather troubling properties of heroin and stopped making it. It was hoped that oxycodone would retain the painkilling effects of heroin without the addictive properties. This turned out not to be the case. During World War II oxycodone was the main battlefield painkiller used by the German army. Adolf Hitler's physician, Dr Theodor Morell, gave Hitler repeat injections of oxycodone and, when it was no longer possible to obtain it, Hitler likely went into full-scale withdrawal in January 1945, which might well explain his behaviour during the latter part of World War II.[22] Oxycodone was also routinely given to U-boat captains to improve their performance. When the Nazi leader Hermann Göring was taken into custody by the Americans, he had thousands of doses of oxycodone in his possession.

In the 1990s the drug company Purdue Pharma developed a prescription version, which they named Oxycontin. When it was launched in 1995 it was hailed as a medical breakthrough that could help patients suffering from moderate to severe pain. The drug became a blockbuster and generated $35 billion in revenue for Purdue. Yet Oxycontin rapidly became a drug of abuse, with many people becoming addicted, both from medical and from illicit use. The latest US data shows that Oxycontin abuse has claimed the lives of a staggering 400,000 people between 1999 and 2017. In the last few years life expectancy in the US has fallen, unlike in most developed countries. One reason for this is the current opioid crisis, with Oxycontin being called the 'jet fuel' of the crisis.

In 2017 the *New Yorker* magazine published an article claiming that Purdue Pharma's founders, Raymond and Arthur Sackler, encouraged business practices and direct pharmaceutical marketing to increase sales, which eventually led to the rise of addiction to opioids in the US.[23] The business practices included organising conferences in resorts in Florida, Arizona and California for more than 5000 physicians, pharmacists and nurses, where all expenses were paid.[24] Purdue repeatedly targeted physicians who were low prescribers. There was a lucrative bonus system for salespeople. In 2001, Purdue paid $40 million to its sales staff in bonuses. Purdue also distributed branded promotion items, such as Oxycontin fishing hats, expensive toys and a compact disc entitled 'Get into the swing with OxyContin'. This promotion campaign also minimised the risk of addiction, which Purdue has admitted to. Remember, Oxycontin has claimed the lives of over 400,000 people. All of this has given rise to litigation against Purdue Pharma, since it is alleged that Purdue Pharma and members of the Sackler family knew that high doses of Oxycontin over long periods would greatly increase the risk of addiction. The Massachusetts Attorney General, Maura Healy, has accused Purdue of deceiving patients and doctors about the addictive and deadly risks of Oxycontin. Massachusetts is one of many jurisdictions that are suing Purdue and, in some cases, the Sackler family themselves for the damage done by Oxycontin. The opioid epidemic in the US is currently killing 200 people per day.

The claim is that Purdue deceived doctors and patients to get 'more and more people on its dangerous drugs' and 'mislead them to use higher and more dangerous doses'.[25] The allegations against Purdue include that they hired many hundreds of sales staff and taught them false claims to use to sell Oxycontin. There are also claims that Purdue engaged in aggressive promotion campaigns, paid selected 'key opinion leaders' to make what appeared to be unbiased endorsements of Oxycontin and targeted its marketing to vulnerable patient groups such as the elderly and veterans. Purdue denies the allegations. There is evidence that scientists in the federal

ARTHUR M. SACKLER (1913–1987),
FOUNDER OF PURDUE PHARMA.
HE DIED BEFORE OXYCONTIN WAS
INTRODUCED.

government, as well as scientists in Purdue, warned Richard Sackler of the risk that Oxycontin would be abused if uncontrolled. Purdue Pharma filed for bankruptcy in September 2019 and has offered $12 billion to settle around 2,000 lawsuits (which have come from 23 states), but this has been rejected and litigation continues. Purdue has stated that it will dedicate all of its remaining assets and resources 'for the benefit of the American public'.

Although chemical addiction is the primary type of addiction, behavioural addiction is becoming more common. Behavioural addiction is a chronic obsession with an activity such as gambling, sex, gaming or use of the internet and smartphones. The internet is seen as especially problematic as it encourages reclusive behaviour and personal isolation, both of which promote behavioural addiction. In Ireland, gambling is common, with 44 per cent of people playing the lotto weekly, 12 per cent placing bets with bookmakers and around 2 per cent gambling online.[26] The Irish bet an incredible €5 billion per year. Gambling is the only form of behavioural addiction recognised by the bible of what constitutes mental health issues – the *Diagnostic and Statistical Manual of Mental Disorders*. But this designation is controversial, and addiction counsellors are often of the view that addiction to sex, gaming and the internet are equally worthy of inclusion.

The list of chemicals or behaviours that lead to addiction would lead us to wonder why everyone isn't addicted to something. But why do some of us succumb to addiction and some of us not? As with other human traits, the answer will lie somewhere along the continuum from genetics to the environment. Understanding more about this critical issue is important for the effort to provide help to people who desperately want to escape their addiction. There is no doubt that genetic factors play an important role, but even those at a low genetic risk who are exposed to high doses of an addictive substance over a period of time will become addicted.[27] For many drugs it seems to be a matter of dosage. There's a well-known statement in the discipline of toxicology: everything we consume is a poison, it's just a matter of dose. With drugs, it looks like some of us have a lower threshold for addiction than others. This might be caused by genetic susceptibility, with a sensitivity in the brain caused by differences in the levels or activity of proteins that are sensing the drugs, such that pathways that lead to addiction are triggered. There is also the well-known feature of tolerance, where the brain tries to protect itself and ramps down the sensing of the drug: drug-users become tolerant to the drug, so they need more to get the same effect and overcome the decreased responsiveness.

Lots of family studies have been done to find the needle in the haystack that points to a genetic susceptibility to addiction.[28] These include studies of identical twins, fraternal twins (who are no more genetically identical than regular siblings, but who are likely to have more similar environments than regular siblings), siblings and adoptees. The needle has to be found in the 0.1 per cent of our DNA that makes each of us unique. When taken together, around half of a person's risk of becoming addicted is based on his or her genetic makeup.[29] With identical twins, if one twin is addicted to a substance, it is rare for the other twin not to be addicted, but this is not the case with fraternal twins. If one family member has an addiction, the chances of another family member developing the same habit are high. When addiction to specific substances has been examined, interesting findings emerge. Around 30 per cent of cannabis users will develop an addiction.[30] A study of 2,387 cases and 48,985 controls identified a gene called CHRNA2 that was linked to cannabis addiction. If you make less of the protein encoded by this gene, you had a higher risk of cannabis addiction. This study was replicated in a subsequent study involving 5,501 cases and 301,041 controls.[31] Why this gene is associated with addiction isn't known, but it is the first strong candidate gene for risk of cannabis addiction.

In another study, a variant in the gene for a protein called SLC6A11 has been associated with risk of nicotine addiction.[32] This protein regulates the level of the inhibitory neurotransmitter GABA, a pathway that is targeted by nicotine. The protein produced by the variant may be more susceptible to targeting by nicotine and people with it may be more susceptible to nicotine addiction. Intriguingly, we might have inherited this gene variant from our Neanderthal cousins, the Stone Age people we met when modern humans migrated from Africa into Europe some 100,000 years ago.[33] Humans had sex with the Neanderthals and we are all descended from the resulting offspring. Some of us still carry this Neanderthal gene. Since Neanderthals didn't smoke (as far as we know) it's not clear what the function of that particular gene would be in Neanderthals, but its presence goes some way to explaining the risk of nicotine addiction.

Genetic factors account for around 50 per cent of the risk of developing alcoholism, while cocaine addiction might be as high as 79 per cent.[34, 35] Genetic factors are substantial when it comes to the chances of someone developing an addiction. Having a parent who has a drug or alcohol addiction thus becomes an important risk factor.[36] One study, involving over 100 scientists, examined addiction to nicotine and alcohol in the amazingly large number of 1.2 million people.[37] They also measured behaviours such as the age when smoking began, the age when smoking stopped, the number of cigarettes per day and drinks per week. They then cross-checked the findings with life events, such as years of education and any diseases suffered. Scientists then correlated these findings with genes linked to addiction. The study illustrates how complex the link to environment and genetics is because the scientists reported over 566 genetic variants that influence the risk of addiction to nicotine or alcohol and associated these with the life events measured. The genes encoded proteins involved in how nerve cells fire in our brains and pointed to the old favourites of dopamine, glutamate and acetylcholine. Overall, the scientists concluded that the risk of addiction is indeed a complex combination of genetic and environmental influences. Three genes, CUL3, PDE4B and PTGER3, emerged as being especially important for the risk of addiction and further work on these could prove especially informative.

As for environmental factors, these are of course seen as important as any genetic factors and in fact possibly more so. If we know the environmental

ADDICTION WILL BE
DRIVEN BY NATURE VIA
NURTURE.

factors, there's a chance we might be able to intervene and alter them to reduce the risk. This is because addiction will emerge as a combination of environment and genetics – it will be driven by nature *via* nurture. People may well have certain genetic variants that put them at risk of addiction, but that will only be revealed if the person is in a particular environment.

A number of environmental influences have been implicated: lack of parental supervision, pressure from peers, drug availability and poverty are all proven risk factors.[38] An extensive study of 900 court cases involving people who had experienced abuse as children revealed they were at great risk of developing substance abuse problems later in life.[39] Adverse childhood experiences, which include mistreatment in childhood, including physical and sexual abuse, are strongly associated with the risk of addiction later in life. Globally, a history of child sexual abuse is estimated to be responsible for 4–5 per cent of alcoholism in men and 7–8 per cent in women.[40] This agrees with several studies indicating that girls are more likely to be at risk from adverse life events than boys.[41] A girl who experiences such events in childhood is at a higher risk of developing an addiction later in life than a boy who is exposed to the same level of childhood stress.

What is also interesting is that the more frequent stressful life events a child experiences, the more likely it is that addiction will develop. This has also been shown to be the case in adults. It seems as if we have a set capacity for a number of stressful life events, and beyond that number we are at risk of mental stress, which might lead to addiction. The timing of the stressful event has also been shown to be influential. The quality of the nurturing a child receives in early childhood is especially impactful and overall maltreatment presents a greater risk of addiction.[42]

Studies in animals support the idea that stress in early life is a major risk factor for later substance abuse. Early-life stress in macaque monkeys and rodents has been shown to promote alcohol or substance abuse later in life in drug self-administration models.[43] One study in the 1970s, carried out by psychologist Bruce Alexander, proved to be especially informative.[44] Dr Alexander showed that if a rat in a cage is offered two water bottles, one filled with water and the other with water containing heroin or cocaine, the rat will repeatedly drink the water containing the drug until it overdoses and dies. Alexander then wondered if the experiment was about the drug or the environment. He repeated the experiment, putting the rats in a 'rat park'.

Here they could roam free, play with rat toys and socialise with other rats. The environment was an enriching one. And guess what? They preferred the regular water. Even when they did go for the drug-laced water, they did so intermittently and never became addicted. The results suggested that a social community protects against drug addiction. This is important as it suggests that a stressful environment or upbringing, and/or not being part of a community, is a major risk factor for heroin or cocaine addiction.

Another important environmental factor in the risk of developing an addiction is age. The evidence suggests that a child's neurological development can be permanently damaged if they are chronically exposed to stressful events, including physical, emotional

BRUCE ALEXANDER, WHO USED RATS TO DEMONSTRATE THAT A HARSH ENVIRONMENT WILL ENCOURAGE THEM TO TAKE HEROIN.

or especially sexual abuse. When the child reaches adolescence, they may turn to addictive substances as a coping mechanism.[45] Adolescence is when vulnerability to addiction appears to be at its highest.[46] This is because the so-called incentive-reward systems in the brain mature ahead of the cognitive control centre. The adolescent will be susceptible to intense reward sensations ahead of having the cognitive ability to control the feelings of reward, increasing the risk of addiction. The reward system therefore has the upper hand. Adolescents are well known to engage in impulsive, risky behaviour that might give rise to addiction. Those who start to drink alcohol at a younger age are more likely to develop alcoholism later. Studies have shown that 16 per cent of alcoholics started drinking before the age of 12.[47]

And so, as a result of a complex interplay between environment and genetics, people become addicted. Once addicted, are there any physical signs of addiction in the person's brain and might it be possible to reverse those changes to relieve the addiction? Putting nicotine, alcohol, cocaine or heroin into your brain will have all kinds of effects. It's possible to image the brains of people with addiction and observe changes.[48] Most addictive drugs reduce receptor levels for dopamine in the brain, which is directly responsible for tolerance. Over time, though, regions outside the dopamine reward centre also change. These brain regions are involved in judgement, decision-making, learning and memory, so all of these functions also become compromised. Stopping drug use doesn't necessarily lead to immediate restoration of these areas. Some drugs actually kill particular neurons that are never replaced. This can make it difficult for addicts to return to the way they were before they became addicted. Dopamine is especially interesting when it comes to addiction. A precursor of dopamine, L-Dopa, is used to treat Parkinson's disease, in which neurons that make dopamine die. These neurons that die are involved in movement, so the key feature of Parkinson's is altered movements (for example, enhanced tremor). L-Dopa boosts dopamine levels and can relieve symptoms, but the problem is that it can give rise to addictive behaviour, with people developing so-called impulse

control disorders such as gambling, overeating and hypersexuality, which is directly caused by the L-Dopa.[49] Since L-Dopa boosts dopamine levels, this strengthens the conclusion that dopamine plays a vital role in addiction.

TRAINSPOTTING (1996) FEATURES THE MEMORABLE LINE FROM LEAD CHARACTER RENTON, 'WE WOULD HAVE INJECTED VITAMIN C IF ONLY THEY HAD MADE IT ILLEGAL!'

How can people escape the clutches of addiction? A person with addiction often has to convince themselves of their own addiction. There can be a lot of denials, especially when it comes to alcoholism and heroin addiction. It's a rare person who will admit to these addictions, for fear of being labelled weak-willed and a failure. It's worth stating what an addiction is. First, you will have an urge to use the drug every day and sometimes multiple times a day. Second, you will become aware that you are taking more drugs than you actually want to, but you can't help yourself. Third, you will try to ensure that you have the drug with you at all times. Fourth, you will buy it even when you know you can't afford it. Fifth, you keep using the drug even though you know it's causing trouble at work or with family and friends. Sixth, you will spend more time on your own, and you won't take care of yourself or be too concerned about your appearance. Seventh, you will do dangerous things like driving while you're on the drug or having unsafe sex. Finally, you will have no compunction about lying or stealing to obtain the drug.

This is a scary enough list of the toll of drug addiction, both on you and the people close to you, who will also suffer. But there is hope. It's important for addicts to know that being addicted to drugs or alcohol is not due to a character flaw or a sign of weakness:[50] it is due to a combination of factors. Studies have shown that the first step towards recovery is the toughest: the person with addiction must recognise they have a problem and decide to make a change.

Recovery usually starts with talking to a doctor, who will know how to help. Detoxification may be needed to clear the body of the drug. People with addiction will have to avoid friends who still use, and also avoid bars or clubs where they used to go because the environment where they drink or take drugs can be a trigger, which will enhance craving. Behavioural counselling will be needed to help them identify root causes, repair their relationships and learn coping skills. Medication may also help. Two examples are the use of methadone for heroin users and various nicotine-replacement approaches for smokers. Methadone is a drug somewhat similar to heroin. It is used to help people wean themselves off heroin. It relieves pain but also blocks the high from heroin. Yet its use has been criticised for not so much curbing addiction as maintaining drug dependency through an authorised, government-sponsored provider. For smokers, nicotine replacement, which involves such products as nicotine patches and chewing gum, reduces the craving for nicotine and increases the chances of quitting smoking by 50–60 per cent.[51] Drugs like Bupropion, an antidepressant, have also been shown to help people stop smoking and it improves the chances of quitting 1.6 fold. E-cigarettes, also known as vaping, have also been shown to help; vaping involves the inhalation of vaporised nicotine and is seen as a real alternative to smoking because it might ultimately lead to smoking cessation. It still has nicotine, but it's seen as the lesser of two evils as it is less dangerous to overall health. A recent study compared vaping to other nicotine replacement approaches in a year-long trial and found that, of 886 participants, 18 per cent in the e-cigarette group had given up smoking, compared to 9 per cent in the nicotine replacement group.[52] Both groups had received behavioural support during the trial.

Addiction remains a major health issue for so many of us, and a major challenge for society at large. It seems to be part and parcel of being a human. **The bottom line: if you are unfortunate enough to have an addiction that is having a negative effect on your life, remember, it's because you're a member of the human race and there is hope. With the right care you can escape your addiction and continue to live a fulfilling life.**

CHAPTER 7

WHY SHOULDN'T DRUGS BE LEGAL?

—

'I loved it when Bush came out and
said we're losing the war against drugs.

You know what that implies?

There's a war being fought, and the
people on drugs are winning it.'

—

Bill Hicks

CONFESSED IN CHAPTER 6 that I had taken ... wait for it ... a drug that is illegal in Ireland: cannabis. Gulp. Will the Garda come knocking? I am not alone. (Isn't that nice?) In 2019 the United Nations Office on Drugs and Crime released the latest on illegal drug use in the world:[1] 271 million people had used illegal drugs in the previous year. This represented a 30 per cent increase since 2009. The report also revealed that the global illicit manufacture of cocaine had reached an all-time high (so to speak). The US government spends around $50 billion per year trying to eradicate drug abuse, resulting in less than 10 per cent of illegal drugs being seized.[2] If drugs were available legally, they could be regulated and taxed. It is estimated that this could generate an income of $58 billion,[3] a net gain of $108 billion for the American government. Meanwhile, in Europe, €30 billion is spent on illegal drugs each year.[4] Legalising drugs is predicted to hugely reduce the crime rate. The fact that drugs are illegal leads to murder, violence and theft. These crimes are widespread and endemic, and one reason why is the illegality of drugs. The war on drugs has been deemed a failure in report after report. The Cato Institute in the US, which examined the issue closely, has stated: 'We conclude that prohibition is not only ineffective, but counterproductive, at achieving the goals of policymakers both domestically and abroad. The war on drugs has contributed to an increase in drug overdoses and fostered and sustained the creation of powerful drug cartels.'[5]

Why then is the use of drugs by adults not legal and controlled? Drugs obviously have dangers and can cause much suffering, so is banning them simply a matter of the lesser of two evils? Or is it literally high time to do something radical to deal with a growing problem? Might a rational grown-up response to this issue actually benefit society in the long run?

WHEN ELVIS PRESLEY VISITED THE WHITE HOUSE TO ASK NIXON TO MAKE HIM A 'FEDERAL AGENT AT LARGE' FOR THE BUREAU OF NARCOTICS AND DANGEROUS DRUGS (PRETTY INAPPROPRIATE, GIVEN THAT ELVIS HAD A DRUG ADDICTION).

The term 'war on drugs' was coined by the US media shortly after a press conference on 18 June 1971 by Tricky Dicky himself, president of the USA, Richard Nixon. He declared drug abuse 'public enemy number one'. His passionate advocacy has since been interpreted as racism.[6] The theory goes that he knew the war on drugs would disproportionately attack black communities. It has continued ever since then, although the term 'war on drugs' is now seen as counterproductive.

Attempts to control recreational drug use have been going on since the nineteenth century. The first modern law in Europe for regulating drugs was

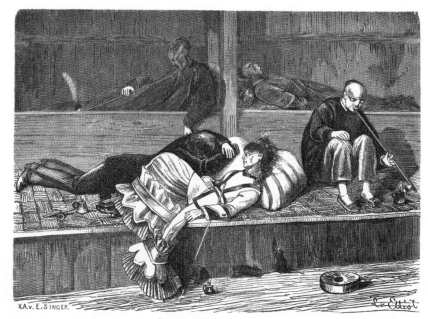

X.A.v. E.SINGER.

San francisco: Opium den.

NO, NOT AN IMAGE OF A TYPICAL STUDENT APARTMENT. AN ENGRAVING OF AN OPIUM DEN FROM THE 19TH CENTURY.

the Pharmacy Act, which was established in the UK in 1868. The act set controls on the distribution of drugs and decreed that opium-based products should only be bought from a seller known to the purchaser and only in sealed containers with the seller's name and address. British traders had long sold opium to the Chinese as a way of balancing trade with China, since there were a lot of Chinese imports coming into the UK, comprising silk, porcelain and tea. Protestant missionaries in China objected to the trade and produced the first-ever scientific analysis on the dangers of opiate addiction in a publication entitled *Opinions of Over 100 Physicians on the Use of Opium in China*.[7] Several nations eventually signed a treaty to control the use of opiates, which was even incorporated into the Treaty of Versailles.

In 1914 a US law was passed to restrict the distribution and use of certain drugs, under the Harrison Narcotic Tax Act. The first evidence for morphine addiction in the US was during the American Civil War, when wounded soldiers were treated with morphine, which was seen as a new-fangled wonder drug. It relieved pain, asthma, headaches and menstrual cramps. Doctors loved it, as it worked quickly and the patient went away happy. The Union army in the Civil War issued almost ten million opium pills to its soldiers.[8] Many soldiers returned from the war addicted. By the 1880s medical journals in the US began reporting extensively on the danger of morphine addiction. A campaign was launched to ban opiates. This was in part driven by the view that smoking opium was an evil practised by Chinese immigrants, but also by gamblers and prostitutes, and so should be stamped out. In 1875 San Francisco banned smoking of opium in Chinese opium-smoking dens because 'many women and young girls were being induced to visit the Chinese opium den, where they were ruined morally'.[9] The law only banned Chinese people from selling opium.

Until 1912 drugs such as heroin were actually sold in over-the-counter cough syrups. Doctors could also prescribe heroin for irritable babies (it certainly calmed them down), insomnia and 'nervous conditions'. Then in 1914 the law banned the use of opiates like opium, morphine and heroin. This was followed in 1919 by the Eighteenth Amendment to the United States Constitution, which banned the sale, manufacture and transportation of alcohol, and so began the era of prohibition. This was repealed in 1933, having failed spectacularly. Heroin and cocaine joined the list of banned drugs in 1920.

In 1937 the Marihuana Tax Act was passed to ban cannabis. It has been hypothesised that the motivation for this was to ban hemp, a strain of the cannabis plant that is grown for industrial uses, which could be used to make various products, including paper and fibre for textiles. The pressure to ban it came from the DuPont corporation, who had invented nylon and wanted to kill off the competition. Andrew Mellon, who was Secretary of the Treasury in the US government, had himself invested heavily in Du Pont because of the promise of nylon, and lobbied for hemp to be banned.[10]

In China, Mao Zedong almost eradicated the opium trade in the 1950s: ten million addicts were forced into compulsory treatment and dealers were executed.[11] But the Vietnam War led to a surge in demand, with 20 per cent of US soldiers regarding themselves as addicts in 1971.[12] By 2003, in spite of severe penalties, China was estimated to have four million regular heroin users.[13]

A STAMP COMMEMORATING 'BICYCLE DAY' IN APRIL 1943, WHEN CHEMIST ALBERT HOFMANN TOOK THE PSYCHEDELIC DRUG LSD AND THEN CYCLED HOME.

From the 1960s onwards the war on drugs was ramped up significantly and LSD joined the list of banned drugs in 1966. LSD has an interesting history: it was first made in a lab in 1938 by chemist Albert Hofmann, who worked for the drug company Sandoz.[14] He was trying to make a drug to stimulate breathing and blood circulation. It took another five years for Hofmann to inadvertently sample some and take the first trip. He then took some on purpose (at ten times the dose needed to have an effect) and had a psychedelic experience, cycling home on his bicycle on what is now known to aficionados of psychedelic drugs as 'Bicycle Day'. He never thought it would be used recreationally and instead felt that it might have uses in psychiatry. In the US in the 1950s LSD was tested on undergraduate psychology students and *Time* magazine published a number of positive reports on LSD use in 1954 and 1959. Psychoanalyst Sidney Cohen became a strong advocate and began promoting the drug as a treatment for alcoholism and also to stimulate creativity. He also gave some to writer Aldous Huxley whose book *The Doors of Perception* was inspired by LSD use (and from which title The Doors got their name). In one trial, LSD was given to alcoholics and showed impressive results: after one year 50 per cent had not had an alcoholic drink.[15]

By the 1960s over 40,000 patients had been given LSD, mainly for psychiatric disorders. Many psychiatrists began taking it recreationally. Then, in 1965, Sandoz stopped making it, because of growing government protests over its use by the general public. In 1966 LSD was deemed illegal in most countries but continued to be used by artists to inspire creativity. To quote comedian Bill Hicks again (the patron saint of liberalising drug laws): 'If you don't believe drugs have done good things for us, do me a favour. Go home tonight. Take all your albums, all your tapes and all your CDs and burn them. Because you know what, the musicians that made all that great music that's enhanced your lives throughout the years were *rrreal* fucking high on drugs. The Beatles were so fucking high they let Ringo sing a few tunes.'

COVER OF THE BEATLES 1966 ALBUM *REVOLVER*. MANY SONGS REFLECT THE BAND'S INTEREST IN LSD.

From 1969 drugs in the US were classified under the Controlled Substances Act.[16] In Ireland, the Misuse of Drugs Act 2015 is the main law concerning drug use.[17] Schedule 1 drugs are deemed to have no medicinal use and to have a high likelihood of being abused. They include cannabis and LSD. Schedule 2 drugs have a high likelihood of being abused but have medicinal use and include cocaine (which can be used as an anaesthetic), the opiates fentanyl, heroin, methadone, oxycodone (which can all be used in pain relief) and amphetamine (which can be used for narcolepsy). Schedule 3 drugs are frequently prescribed to the public for common ailments, but also carry the risk of abuse and comprise many prescribed drugs. Schedule 4 drugs contain a wide range of medicinal products with limited risk of abuse. In 2010, in response to the threat posed by new psychoactive substances being sold in so-called head shops, 200 new substances were classed as illegal. New drugs are invented all the time, sometimes making it difficult for legislation to keep up. In the past ten years, 700 new psychoactive substances have been made which mimic cannabis or cocaine.[18]

In Ireland, possession of drugs without a prescription that falls under the Misuse of Drugs Act is illegal and carries various penalties under the law. In

2017 there were 12,211 offences for use or possession and 4,175 offences for supply.[19] The penalties applied vary depending on the drug. The majority of European countries carry a penalty of incarceration (as opposed to just a fine) for possession. Spain, Italy and Switzerland do not impose a custodial sentence for minor drug possession for any drug. Drug trafficking carries the highest penalty in Europe as a whole. In Ireland, possession of drugs with a market value of at least €13,000 carries a ten-year sentence.

The legal situation of cannabis, in Ireland and globally, is interesting. Laws are changing and overall there is a general loosening up when it comes to cannabis. In Ireland it's still illegal, and possession carries a penalty of up to €1,000 for a first offence and €2,540 for a second offence. Third and subsequent offences are punishable with a sentence of up to three years. Possession of any other illegal drug carries a prison sentence of up to one year for a first offence or a community service order. The legality of cannabis for medical and recreational use varies by country in terms of possession, distribution and cultivation. The use of cannabis recreationally in most countries, including Ireland, is illegal, but many countries have decriminalised it. Asian and Middle Eastern countries still have severe penalties for possession. Four countries have legalised its recreational use: Canada, Georgia, South Africa and Uruguay. To many people's surprise, given how puritanical many Americans are, 11 states in the US have made cannabis legal, although federally it remains illegal. A policy of 'limited enforcement' has been adopted in Spain and the Netherlands, where sale of cannabis is tolerated in licensed establishments.

Why was cannabis legalised in the US, at least in some states? From the late twentieth century, there was a move towards legalising cannabis. In 1996 California relented first, decriminalising it for medical use (which effectively meant anyone could use it if they saw fit). Then, in 2012, Washington and Colorado legalised recreational use. By early 2019 more than 30 states permitted some use: then, on 20 November 2019, a judiciary committee in the House of Representatives passed a landmark bill, legalising cannabis at

A 1972 POSTER ADVERTISING A 1936 MOVIE *REEFER MADNESS*, WHICH WARNED AGAINST THE DANGERS OF MARIJUANA AND GAINED A CULT STATUS IN LATER YEARS.

the federal level and removing it from Schedule 1 of the Controlled Substance Act. The bill has yet to be passed in the Senate, and there is concern that it might be blocked, since Republican majority leader Mitch McConnell opposes cannabis legislation.

The ongoing debate on legalisation in the US reopened the question as to why cannabis was made illegal in the first place. As with Nixon's war on drugs, the uncomfortable answer that emerged was racism.[20] Close analysis revealed that in the early twentieth century cannabis was hardly used at all. Then Mexican immigrants brought cannabis with them and fears began to be stoked that it created a 'lust for blood' in immigrants – the term 'cannabis' was then increasingly replaced with 'marijuana', which was a way to convey its foreignness and promote xenophobia. Marijuana and cannabis are used interchangeably, although cannabis generally refers to the actual plant and marijuana refers to a drug preparation from it. The word marijuana originated in Mexican Spanish. It may have come from 'marjoram' (another herb) or from the name Mary Jane. There are of course many names for cannabis or drug preparations from it, including pot, weed, dope, grass, herb, skunk and ganja. One has to wonder if cannabis users are like the Irish, who rated seaweed so highly that they have an estimated 31 words for it.

In the 1920s many states banned cannabis. In the 1930s Harry Anslinger, the then head of the Federal Bureau of Narcotics, imposed a federal ban in spite of the majority of scientists who were surveyed at that time stating that cannabis was not dangerous. Anslinger claimed the majority of cannabis users were African Americans and that it had a negative effect on what he called 'degenerate races'. He is on record as saying 'Reefer makes darkies think they're as good as white men.' Anslinger was also of the view that smoking cannabis would result in white women having sex with black men. Remember, he was Director of the Federal Bureau of Narcotics. By the time we get to the early twenty-first century, African Americans were shown to be almost four times more likely than Caucasians to be arrested for cannabis possession, despite both groups having similar usage rates.[21] According to the American Civil Liberties Union, marijuana arrests currently account for more than half of all drug arrests in the US.[22] More importantly, the impact on minority communities is much greater than others. And there is the three-strikes rule: the third time you're caught with cannabis, you receive a mandatory life sentence. Again, this disproportionately affects minority communities, particularly African Americans. Arguments to protect these minorities were put forward during the debate, and the bill that was passed also allowed for a 5 per cent tax to be put on sales of cannabis. Slowly, the US is moving towards legalising cannabis for reasons to do with its overall safety profile, the racist nature of the legalisation, the burden its illegality places on policing , as well as the costs involved in keeping offenders in prison. And of course, the chance to tax it and raise revenue.

Legislators in Europe are closely watching what's been happening in the US in relation to cannabis. Many countries have legalised the medical use of cannabis: cannabis or its extracts bring substantial benefits, including relief of nausea and vomiting during chemotherapy, improved appetite for people with AIDS and other diseases, and pain and spasm relief in MS. As regards legalising it for recreational use, unlike the US, no European country (with the exception of Switzerland) has provisions that allow voters to directly change laws, although referendums can be held to advise lawmakers. Nearly all the changes that have happened in the US have come about as a result of popular votes in different states. For Europe, lobbying of politicians is the main method of bringing about change, as well as providing the media with evidence to support change. Perhaps Europe will always lag behind the US when it comes to cannabis. It was the US who first made it illegal, and this spread to Europe. The US may be the first country to lift the ban, and Europe may then follow suit.

Other countries employ a variety of approaches to drug legalisation. The Dutch have always taken a pragmatic view of what other countries called 'vice'. Prostitution has been legal and regulated in the Netherlands for decades and has been subjected to taxation since 2011. In the twentieth century the Dutch were of the view that a drug-free Dutch society was unrealistic and impossible to achieve, and that the effort should be put into minimising harm.[23] They adopted a policy of *gedoogbeleid* meaning 'tolerance policy'. We all need some *gedoogbeleid* in our lives. Distinction is made between 'soft drugs', which include cannabis, sleeping pills and sedatives, and 'hard drugs', which include heroin, cocaine, amphetamine, LSD and MDMA. Soft drugs are tolerated whereas hard drugs are definitely illegal. So-called Coffeeshops (one word, to distinguish them from regular cafés) are allowed sell soft drugs. This somewhat lax attitude has had an unintended consequence: drug trafficking in the Netherlands is rampant. The country is a major transit point for drugs entering Europe, including cannabis, heroin, cocaine and amphetamines. Despite efforts from Interpol, the Netherlands in the late 1990s became a major exporter of the sedative temazepam.[24] Laws were passed in 2005 that make it illegal for coffeeshops to sell cannabis to non-Dutch people, to prevent cannabis users from other countries flocking to the Netherlands.

A study into the level of cannabis, cocaine, MDMA and methamphetamine (or by-products of these) in waste water revealed that out of 42 major cities, Amsterdam came near the top of the list.[25] 'Notorious party cities' as the report said, were all in the top ten. The Serbian city of Novi Sad topped the list for cannabis, with Amsterdam second and Paris coming in third. Cocaine levels were high in Amsterdam, Antwerp, London and Zurich (bankers like cocaine …). MDMA levels were highest in Amsterdam and in Switzerland, with MDMA spiking on Sundays, presumably because of weekend use. The Netherlands spends more that €130m annually on facilities for people with drug addiction. Its so-called 'demand reduction programme' reaches 90 per cent of the country's hard-drug users, the number of which has stabilised in recent years with the average user age rising to 38, indicating that young people are less likely to use hard drugs. Might the Netherlands be an example for the rest of the world to follow?

Australia is another interesting country to consider when it comes to drug policies.[26] In 1985 it came up with a National Drug Strategy, whose aim was

to move drug use from prohibition to harm reduction, in addition to demand reduction (prevention and treatment) and supply reduction (customs and policing). But when examined closely it turns out that only 2 per cent of funding goes towards harm reduction, while 66 per cent goes to law enforcement. Some 30 years ago Australia removed criminal penalties for possessing cannabis for personal use. The ban on MDMA in Australia led to its illegal manufacture, with the subsequent preparations having unknown potency and contaminants of unknown toxicity.[27] The numbers in Australia in relation to crime continue to stoke debate on what to do about the issue; in 2016–17 there were 113,533 illicit drug seizures and 154,650 drug-related arrests.[28]

THE DUTCH HAVE ALWAYS BEEN OF THE VIEW THAT A DRUG-FREE SOCIETY IS UNREALISTIC AND IMPOSSIBLE TO ACHIEVE. SOFT DRUGS CAN BE BOUGHT BY DUTCH CITIZENS IN SO-CALLED 'COFFEESHOPS'.

The Portuguese have perhaps the most enlightened system when it comes to illegal drugs, and policies there appear to be working.[29] In 2001 a new law was enacted, which maintained the status of illegality for use or possession for personal use. However, the offence was no longer criminal, meaning it may not lead to a jail sentence for those found guilty. A major expansion in harm-reduction efforts followed, with the level of public investment in drug treatment and drug prevention doubling. Has this shift in policy worked? Numbers suggest that it has. There has been a significant reduction in drug-related deaths. Drug use among adolescents and 'problematic' users has declined and HIV rates are down, presumably because of needle-exchange programmes and there being fewer heroin users.[30] Drug-related criminal justice workloads have also decreased.

An obvious and major reason as to why drugs remain illegal in most countries is concern about the risk to health and well-being, especially in young people. In 2010 a group of experts ranked 20 legal and illegal drugs on 16 measures of harm to the user themselves, and to the wider society.[31] Alcohol was easily the most harmful, coming in with a score of over 70/100: next came heroin at 55/100, followed by crack cocaine at 53/100, metham-phetamine (also called crystal meth) at 32/100 and cocaine at 27/100. Next came tobacco at 26/100 and amphetamine at 23/100. Cannabis was next at 20/100. MDMA (also called ecstasy), LSD and mushrooms all scored below 10/100. Based on these criteria, alcohol should definitely be made illegal for everyone (not just the sale of it to minors) – although we have seen how that can go badly wrong during Prohibition in the US.

Detractors of the legalisation of drugs raise a number of valid concerns.[32] It might encourage experimentation in those who have a predisposition to addiction. Prices will fall, which might lead to increased use. It's likely that there would be a need for more treatment centres for people with addiction, which will come at a financial cost. But there is also a list of advantages to society should drugs be legalised, which includes an actual decrease in addiction and substance-abuse rates, as seen in Portugal. One reason for this is that, because people with drug addiction aren't being jailed, they are treated more effectively and as a result there are higher rates of recovery. Legalisation also allows people who are addicts and who recover to remain in society – they can gain meaningful employment, have less of a need to turn to drugs and are no longer labelled as criminals. Also, when drugs are illegal, a counter-culture arises that actually celebrates drug use.

Most importantly, there is a strong argument that legalisation would allow the criminal justice system to focus on what it does best: keeping the general public free from harm. Drug enforcement takes up vast amounts of resources in many countries. Obviously, laws against drug use are designed to stop people from using harmful substances, but advocates of reform are of the view that this would be better served through counselling and treatment facilities. Where might the money come from to support such facilities, but also educate the public on the dangers of drug use? Substantial savings will be made from policing and revenue will be generated from taxing drugs. These funds can be reinvested in programmes to help people with addiction but also provide information to the general public on the dangers of drug misuse (including alcohol). Regulation of the drugs being sold would ensure that they are safer, free of toxic contaminants. It all looks like a no-brainer, doesn't it?

Perhaps the biggest concern from drug legalisation is that children and teenagers would have more access to them. This of course also applies to alcohol (which teenagers have no trouble in getting, it seems). Drugs are especially dangerous to the developing brain. Because their brains are still developing, probably up to the age of 25, young people are at a higher risk of addiction than adults.[33] Drugs have a much more intense effect on the younger brain, than on people who are over 25. Addiction to drugs or alcohol can slow brain development, with effects evident in the frontal cortex and limbic systems.[34] Studies in animals have shown that brain circuits which affect decision-making can be particularly affected, especially in response to cannabis, cocaine and MDMA.[35] In humans, chronic use of MDMA has been shown to have toxic effects on several brain regions,[36] which might lead to problems with concentration and mood stability. There is also evidence that long-term use of MDMA decreases empathy.[37]

The changes may be irreversible. Drug or alcohol use can lead to a variety of mental health disorders including depression, personality disorders

HOW A FEW PILLS CAN AGE YOUR BRAIN BY 40 YEARS.

Loss of memory, depression, anxiety, brain injury and scarring are all signs that your brain is ageing, and new research out of Royal Perth Hospital suggests that these are also problems commonly shared by many ecstasy users. See the damage even one pill can do to your brain, as well as a video chat on the new research, at **drugaware.com.au**

DRUG AWARE

AN AUSTRALIAN POSTER
WARNING OF THE DANGERS
OF TAKING ECSTASY. WHO
WOULD WANT 'LOSS OF
MEMORY, DEPRESSION,
ANXIETY, BRAIN INJURY AND
SCARRING'?

and even psychosis. One study concluded that among 23,317 young people studied, cannabis use was associated with an increased risk of depression, although the increased risk was relatively low at 7 per cent.[38] This didn't mean cannabis promotes depression; it was an association. Yet given the numbers who use cannabis, this represents a relatively large number of people. Teenage brains also adapt more quickly to repeated drug use, leading to higher levels of craving and dependence. Ninety per cent of people with drug or alcohol addiction begin substance abuse before the age of 18,[39] although there is only limited evidence that starting on cannabis can lead to using harder drugs (the so-called 'gateway' effect).[40] Given that campaigns to encourage teenagers to say 'no' to drugs have failed, the advice now is to tell them to say 'not yet' and wait until they are older.

Whatever about safety aspects, another concern is that legalisation will increase use among teenagers; yet several recent reports have found that in the majority of states that have approved use of recreational cannabis, use among teenagers actually decreased.[41] One study showed that there was a drop of 8 per cent in use in teenagers who said they had used cannabis in the past 30 days. The study analysed data from 1993 to 2017 in 1.4 million high school students. The reasons for this may be a reduction in the 'forbidden fruit' effect as well as decreased access as cannabis moved from the streets to licensed dispensaries where you need to be over 21 to purchase it. Another study has recently shown that the rate of 'cannabis use disorder' has risen in people over 26.[42] Researchers examined surveys completed by 505,796 people between 2008 and 2016 in states where cannabis was legalised – Colorado, Washington, Alaska and Oregon. Cannabis use disorder is where a person's use of cannabis affects their lives in a negative way – for example, they find it hard to cut down, or cannabis affects their working life or relationships. Problematic cannabis use rose, from 0.9 per cent to 1.23 per cent. Not a massive number, but still significant for those affected. There was no increase in those aged between 18 and 25. The issue appears to be one of heavy use, as it is well known that this is linked to a range of problems, with long-term

health, economic and social consequences. The authors concluded that
the study shouldn't imply that cannabis remain illegal. Legalisation efforts
should coincide with educational measures to promote prevention as well as
support for problem users.

It seems highly likely that cannabis will be legalised in many other
countries, and available to adults, just like alcohol, as is already the case in
many US states. But what about other drugs? There are clearly pros and cons
to legalising drugs, whether they be soft or hard. The case for legalising hard
drugs such as heroin seems a long way off, given the risk of addiction: no
one wants to see a continuation of the opioid crisis in America, which was
started by a prescribed opiate, Oxycontin. If heroin were to be legalised the
situation would be different as there are now strict controls on the use of
Oxycontin. Should we turn again to Bill Hicks, who said: 'What business is
it of yours what I do, read, buy, see, or take into my body as long as I do not
harm another human being on this planet? And for those who are having a
little moral dilemma in your head about how to answer that question, I'll
answer it for you. NONE of your fucking business. It's not a war on drugs, it's
a war on personal freedom.' Do we need laws to protect us from ourselves?

Rich and famous people have always been able to get whatever drugs they
want, and it often ended badly. When Elvis died at the criminally young
age of 42, a post-mortem revealed that he had extremely high levels of four
prescription opiates in his blood.[43] Constipation is a common side effect
of opiate use and sadly Elvis died of a heart attack while on the toilet. His
personal physician, 'Dr Nick', began prescribing opiates for Elvis in 1967. He
said he gave Elvis what he wanted to stop him turning to the street. This
information wasn't released until some years after his death, as it was seen as
something of an embarrassment since Nixon had given Elvis a badge making
him an honorary 'Federal Agent at Large' for the Bureau of Narcotics and
Dangerous Drugs. At the time of his death, Michael Jackson had six drugs in
his system, the most damaging being Propofol: this is only supposed to be
used in hospitals as a general anaesthetic during surgery.[44] Jackson had used

an anaesthesiologist while on tour in Germany in 1996 and 1997 to 'take him down' at night with Propofol, and 'bring him back up' in the morning, because of chronic insomnia. Jackson's insomnia returned most likely because of a commitment he made to play 100 shows in London to help pay off his debts. His doctor, Conrad Murray, administered the fatal dose on the night Jackson died. He was found guilty of involuntary manslaughter and jailed for four years. The other drugs in his body included the benzodiazepine alprazolam (for anxiety) and the antidepressant sertraline.

It's likely that Prince was addicted to opioids for chronic pain and died of an accidental overdose of the opiate fentanyl.[45] This is a powerful painkiller, usually used in patients with severe chronic pain. It's not known where he obtained it, but it's likely to have been from an illegal source. Amy Winehouse died of alcoholic poisoning, with a blood alcohol level five times over the legal driving limit.[46] All these people were addicted to various substances, both legal and illegal, or at least prescribed contrary to guidelines.

How many of us would be like them if we could? How many of us would succumb if drugs were freely available? We can obtain alcohol freely, but would the easy availability of other drugs lead some of us to unwise decisions in relation to their use? The debate on this issue will continue. Clearly illegality has a limited impact on use by the 43 million people using marijuana, 5.5 million people using cocaine, 2.5 million people using ecstasy and almost 1 million using heroin in the US in 2018.[47] **The bottom line: perhaps one day we will reach a point where society will have obtained a level of maturity where we can make up our own minds as to what we put in our bodies, with safeguards and supports in place to protect the young and those who are vulnerable to addiction. How likely is that, I wonder?**

CHAPTER 8

WHY
AREN'T YOU
IN JAIL?

—

'If you're low-income in the United States,
you have a higher chance of going to jail than
you do of getting a four-year degree. And that
doesn't seem entirely fair.'

—

Bill Gates

N APRIL 2019 I was invited into Mountjoy Prison in Dublin, one of Ireland's largest prisons, to give a talk to some prisoners who had read and enjoyed my book *Humanology*. The day before I was due to go in, Anne Keenan, the teacher who had invited me, reminded me not to bring my mobile phone or laptop as they would be kept by security on the way in. She wrote: *Hope you're not scared coming in to visit us!* I emailed back: *I am now!* I spent three hours in the jail, talking to around a hundred prisoners about the origin of life, how humans evolved, what makes us human and where we might be going as a species. I told them that the earth was 4.5 billion years old and said that this was a very long time. A prisoner shouted out 'Not as long as three years in here!' Of all the things I did regarding *Humanology*, this was the most vivid and rewarding. I had a tremendous time, was heckled constantly and got a tiny sense of what it's like to be in prison.

On my way out I asked the teacher what the men who had come to my talk were in jail for. She didn't specify but said something interesting. She said some of them had committed serious crimes but that the vast majority were 'just like you'. This struck a chord. Why was it me standing up there giving a talk to them about my book, as opposed to one of them standing in front of me, sitting there as a prisoner? Why do some of us commit crime and some of us do not? And what can we do about crime in our society? As discussed in the last chapter, will making drugs legal and controlled hugely decrease the crime rate? Will we ever live in a world free of crime?

People are sent to jail because they commit a crime and are found guilty in a court of law. Crime is defined as an unlawful act punishable by a state or other authority. Definitions are important when it comes to the law. Something is a crime if designated as such by the law. The term 'criminal offence' is also used, which is an act that is harmful, either to an individual but also a community, society or the state itself. The idea of having laws to govern society goes back a long time. It most likely began when tribes grew in size and laws were needed to regulate behaviour. The history of law is closely linked to the development of civilisation. Ancient Egyptians had a civil code, divided into 12 books. Sumerians were the first to systematically write down laws over 4,000 years ago. The Old Testament dates back to 1280 BC and was full of laws, the most important being the ten commandments, which covered the main crimes at that time. A famous joke goes: Moses was summoned by God to receive the laws. He was gone for some time and finally arrived back, carrying two large tablets of stone on which were carved the ten commandments. He said to his people: 'I have good news and bad news. The good news is, I got Him down to ten. The bad news is … adultery is still in.'

All religions describe sin, which is a crime against divine law – a law provided by whichever god you believe in to make sure humans behave themselves. Experts on the origin of religion have suggested that when tribes were small (say 100 people or fewer), the leader (or elder) could keep control of things. Once the tribe grew in number, however, an all-seeing supernatural being (perhaps originally consisting of elders who had died) was needed to keep an eye on you and make sure you behaved. This might prevent you from committing a crime. It is possible that crime becomes a feature of society when it reaches a certain number of people. You might be more likely to commit a crime against someone you don't know, and laws may in part have been configured to prevent this. But if you did commit

AN ENGRAVING FROM 1879 DEPICTING THE SEVEN 'CAPITAL' SINS. FROM THE TOP MOVING RIGHT: PRIDE, GREED, LUST, ENVY, GLUTTONY, WRATH AND SLOTH. ENVY IS DEPICTED AS WRITING PAMPHLETS – SOCIAL MEDIA TROLLS, BE WARNED. YOU ARE SINNING!

a crime, then the law would come in to play. Many civilisations provide laws and codes of conduct. Apart from the ten commandments, another well-known list is the seven deadly sins. This became part of Christian teaching and defined behaviours or habits that might directly give rise to crime or sin. The seven deadly sins are pride, greed, lust, envy, gluttony, wrath and sloth. Any one of these might make you commit a crime, although I'm not quite sure how sloth fits in – surely you are too lazy to do much? Maybe you're too lazy to pay your tax return, which constitutes a crime.

Laws are made to ensure people behave themselves in society. If they step out of line and break the law, punishment must be meted out. Laws fall under different categories including offence against the person, violent offences, sexual offences, offences against property, offences against the state, forgery, use of drugs designated as illegal, public order offences and financial offences. Some crimes involve actual harm to others (such as assault), while others are about compliance to decrease the risk of harming others (such as traffic offences). For someone to be found guilty of a crime, there must be either an admission of guilt or presentation of evidence. This is where science comes in: science is all about providing evidence, and this also applies to evidence that a crime has been committed. Forensics involves scientific tests or techniques used for crime detection. Sometimes a decision on whether someone is guilty has to be made 'beyond reasonable doubt'. Forensic evidence can prove problematic, which happened in the case of the Birmingham Six. In 1975 these six Irishmen were each sentenced to life imprisonment following false convictions for bombing two public houses in Birmingham, which killed 21 people and injured 220 others. The true perpetrators were likely to be members of the Provisional IRA, the Irish paramilitary organisation that used terrorism to fight British rule in Northern Ireland. During the trial, scientist Frank Skuse presented forensic evidence that two of the men had handled explosives. In a subsequent appeal, new scientific evidence was presented, which placed 'grave doubt on Dr Skuse's evidence'; the chemicals detected on the hands of the men could have been from other sources, such as playing cards.[1] This was part of the case that led to the men being released after 16 years in jail.

Some countries have laws built on religious beliefs. Abortion was illegal in Ireland because it went against Catholic teaching. A major function of the law is to maintain social order, which basically means protecting others from harm or lack of consideration. A government may also impose laws as a means of social control. If a government wants to force people to behave in a certain way, it will pass a law to make this happen. Once cars were invented, for example, a whole set of laws were needed to regulate behaviour. In 2004, Ireland, to the rest of the world's surprise, passed a law banning smoking in the workplace: the goal was to protect people from second-hand smoke. Given Ireland's well-known pub culture, many were amazed that this would stick, and many publicans were up in arms over the ban. To make sure people complied, a fine of €3,000 was set for those who transgressed. Fines exploit the fact that many worry about their pocket more than they do about harming others. There were exemptions, notably in prisons, but only in a prisoner's cell or in the exercise yard, which were deemed similar to the prisoner's home – the ban doesn't extend to smoking in the home. A total ban on smoking in prisons might have created tension in an already tense environment. This has been an extremely successful law, with only a handful of convictions and fines each year. It is seen as a great success in terms of health,[2] with people being protected from the damaging effects of passive smoking – almost as dangerous as smoking itself.

This brings us to the question of why people defy the law and commit a crime. If someone is brought up to be able to tell the difference between right and wrong, why would they defy this and commit a crime? Whether humans are born with a moral compass or whether they learn this from their parents is still a matter of debate among psychologists and philosophers. Current evidence suggests that babies are in fact born with an innate sense of morality, which parents and society can help develop. Scientists at Yale University performed a recent study that gave a fascinating insight into this question.[3] They studied babies at 5 months of age to see how much they could appreciate the difference between good and bad behaviour. They began with a puppet show. In the show, a grey cat is seen trying to open a plastic box.

The cat tries and tries but fails. A rabbit then appears in a green T-shirt and helps the cat open the box. The scene is repeated, but then a rabbit comes along in an orange T-shirt and closes the box and runs off. The green rabbit is helpful and the orange rabbit is mean. The baby is then presented with the two rabbits from the show. A staff member who doesn't know which rabbit is mean or nice presents the rabbits to the baby at the same time. Just under 75 per cent of the babies prefer the good rabbit. It's pretty amazing that a five-month old can tell the good rabbit from the mean one. Babies seem to have an inbuilt sense of justice.

In ancient times, the explanation for why someone would commit a crime was known as demonology: this is the idea that criminal behaviour is the result of the person being possessed in some way and obviously has its origins in superstition and religion. The first scientific attempt to explain why someone might commit a crime was in 1876, when Cesare Lombroso, an Italian criminologist, came up with the theory of 'anthropological determinism'.[4] (It should be a crime to come up with long words to explain something to boast how smart you are. I would never do that because I suffer from sesquipedalophobia.) This stated that criminal behaviour was inherited and that there were 'born criminals' who could be identified from physical features alone. According to Lombroso, these included large jaws, low sloping foreheads, handle-shaped ears, a hawk-like nose and long arms – and that was just the men. Lombroso also studied female criminals and concluded that they showed fewer signs of 'degeneration' because they had 'evolved less than men due to the inactive nature of their lives'. He was of the

CESARE LOMBROSO (1835–1909), THE ITALIAN CRIMINOLOGIST WHO CAME UP WITH THE TERM 'ANTHROPOLOGICAL DETERMINISM', THE COINING OF WHICH SHOULD HAVE BEEN A CRIME IN ITSELF.

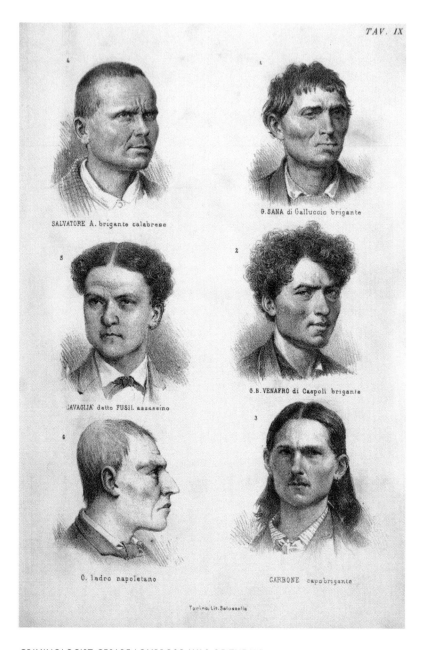

CRIMINOLOGIST CESARE LOMBROSO WAS OF THE VIEW
THAT THE SHAPE OF SOMEONE'S FACE CORRELATED
WITH DIFFERENT CRIMINAL TENDENCIES. CAN YOU
GUESS WHAT HEINOUS CRIMES EACH OF THESE
CRIMINALS COMMITTED? ONE IS OBVIOUS: HAVING A
MULLET (GUY ON THE BOTTOM RIGHT).

view that women weren't intelligent enough to be criminals. Some man, our Lombroso.

Inevitably, Sigmund Freud also weighed in with his opinion. He said that deviant behaviour (meaning deviation away from social norms and the law) came from an excessive feeling of guilt as a result of an overdeveloped superego.[5] He argued that criminals carry large amounts of guilt and commit crimes in order to be punished. Freud was also of the view that humans respond to 'the pleasure principle'.[6] They need to get pleasure from things like food and sex, and if they can't get them, they will commit crimes to obtain them. He was of the view that these urges can be controlled in childhood, and that if such instruction is missing because of poor parenting, the child would grow up into being less able to control these natural urges, and will be more likely to commit crimes to satisfy the needs of the pleasure principle.

Lots of sociologists, psychologists and neuroscientists have since joined the debate on why some people commit crimes and others don't. Sadly, the science behind what they conclude is often badly presented or missing. The area has also been dogged with bias – as we can see in the conclusions of Lombroso, who was clearly a misogynist – and racism. They have, however, managed to clearly define the problem. It is now widely accepted that the reason why someone commits a crime is rarely down to one thing, but is rather a combination of factors, making it difficult to unravel.

The statistics of who is in prison for what crime can be revealing when it comes to trying to understand why people commit crimes in the first place. In 2019 there were 3,996 people in jail in Ireland. This gives a rate of imprisonment of 81 per 100,000. This is pretty average. The USA has one of the highest rates, at over 500 per 100,000.[7] It costs the Irish state an average of €73,802 per year per prisoner. Prisoners serving sentences of fewer than 12 months amounted to 3,559. The majority of people in prison in Ireland have never sat a state examination, and over 50 per cent left school before the age of 15. From 1996 to 2017 the numbers in custody rose by 68 per cent. The average number of females per annum was 165, which fits with international norms. Prisoners are 23 times more likely to come from a deprived area than an affluent one. Around 20 per cent cannot read or write and 30 per cent can only sign their names.[8]

From these kinds of numbers, trends begin to emerge as to who is likely to commit a crime. The first thing that stands out is that a lot more men commit crimes than women. There may be several reasons for this, including social or cultural factors, crimes not being reported and biological factors, such as higher testosterone, which might lead to aggression. The US has some of the most extensive analysis of the gender issue in crime. As many as 14 times more men are in jail in the US than women.[9] In 2014, 73 per cent of people arrested in the US were men, with 80.4 per cent arrests for violent crime and 62.9 per cent arrests for property crime being men.[10] In Ireland, most of the women in jail are there for petty crime; 95 per cent of them are incarcerated for shoplifting or handling stolen goods. As well as committing a crime, men are also much more likely to be victims of crime. In an international 2013 study,[11] 78 per cent of homicide victims were male with the perpetrator being male in 96 per cent of cases. In Ireland in 2018, 77 per cent of homicide and 59 per cent of assault victims were male, but the majority of victims of sexual violence (82 per cent) were women.

The first clue as to why men commit more crimes than females begins in childhood. Boys are much more likely to be delinquent than girls.[12] Studies have shown that overall girls are less likely than boys to have learning difficulties and behavioural problems in childhood.[13] These may set boys on a different life course when compared to girls. 'Life-course-persistent' antisocial behaviour originates in early life and is strongly exacerbated by a high-risk social background, which mainly constitutes a lack of parental support. Testosterone may play a part in making men more aggressive: levels in prisoners were highest in the most violent criminals.[14] Testosterone will drive males to be more competitive, obtain more resources and seek a mate. This, in turn, can lead to crime, including theft and violence; crime might be an extreme form of adaptation whereby a man commits a crime to obtain resources and status. Inter-male competition for resources and a mate may also play a role here. There is a correlation between delinquency and fathering of a child at a younger age.[15] When it comes to aggression, many studies show that men are much more likely to use verbal and physical aggression. Interestingly, an analysis comprising 122 separate studies also found that men are much more likely to be cyberbullies than women.[16]

A final reason for the crime gender gap might be economic. The high crime rate among young men might in part be due to the fact that adverse labour-market opportunities lead to men working in low-paid jobs and as a result, these young men may be swayed by opportunities to commit economically advantageous crimes.[17] Several studies have shown that as the rate of unemployment increases, so does the crime rate – this opportunism may apply to women as well.

We're still left with the question as to why one person, be they a man or a woman, will commit a crime and another not. As with most complex traits in humans, the answer will lie somewhere on the continuum between nature and nurture. In terms of nurture, several things stand out from multiple studies that lead to a person ultimately becoming a criminal.[18] Fear of punishment or rejection keeps most of us from bad behaviour. During childhood, most people take on board society's standard of conduct and will feel guilt, shame and low self-esteem if they commit a crime. For those who commit crimes, several environmental features stand out. The first is antisocial values, also known as criminal thinking. This can happen because of peer pressure, which is itself an important risk factor for someone committing a crime. Belonging to a gang is considered a predicator of future criminal behaviour. Teenagers in particular are prone to peer pressure. Witnessing violence as a child can also have a desensitising effect. A key risk factor is a dysfunctional family, which can mean neglect or an inability to express emotions or communicate effectively in the family environment or at its worse physical or sexual abuse. Someone who has been rejected by their family can find great succour in a criminal gang. These are all environmental factors that can lead to criminal behaviour. But as stated above, it is highly likely that the environment will

RONALD 'RONNIE' KRAY (1933–1995) AND REGINALD
'REGGIE' KRAY (1933–2000) WERE IDENTICAL TWIN
BROTHERS FROM LONDON'S EAST END. WITH THEIR GANG
(KNOWN AS 'THE FIRM') THEY WERE INVOLVED IN ARMED
ROBBERY, ARSON, PROTECTION RACKETS AND MURDER IN
THE 1950S AND 1960S. THEY WERE GIVEN LIFE SENTENCES
IN 1969. (ACCORDING TO MONTY PYTHON, HARRY
'SNAPPER' ORGANS EVENTUALLY CAUGHT THEM.)

combine with underlying genetics to give rise to criminal behaviour.

This leads to the question as to what the genetic basis, if any, to criminal behaviour might be? Are people born under a bad sign? This has been the subject of much analysis,[19] including major research into twins. There are studies comparing the criminal behaviour of identical and fraternal twins. Identical twins will have the exact same genetic make-up, whereas fraternal twins are like regular siblings. If the crime rate between identical twins is the same as between fraternal twins, then it's likely that environmental factors are important, since both sets of twins will have grown up in highly similar environments. If, however, the crime rate is higher between identical twins then fraternal twins, then that would suggest genetics plays a bigger role. Even better if identical twins are separated at birth and brought up in a different environment. If they end up having the same criminal tendencies, then again genetics will trump environment as a causative factor. Overall, the current bottom line is that twin studies support the notion that genetics is an important determining factor in who will become a criminal. This is similar to the risk of devloping an addiction, as we discovered in Chapter 6.

What does the evidence look like? In a Danish study of 3,586 twin pairs (which is a large number for these kinds of studies and so probably gives an accurate result), there was a 52 per cent concordance between identical twins in terms of being a criminal, whereas in fraternal twins this only stood at 22 per cent.[20] The only issue with this and other studies is the possibility that identical twins share an environment more closely than fraternal twins for whatever reason. Overall, though, twin studies support the notion of a strong genetic component in criminality. Another study looked at 31 sets of identical twins and one set of triplets who were all raised apart.[21] Again, evidence was found for concordance between identical twins raised apart and criminality.

Another approach has been to examine children who are adopted. These studies are somewhat more feasible given the rate of adoption in many countries. The first study of this kind was carried out in Iowa and involved

52 adoptees who had been born to women in prison.[22] A group of control adoptees whose mothers weren't in prison, and who were matched for age, sex, race and time of adoption, were also studied. Seven of the 52 adoptees had a criminal record as adults, whereas only one of the control adoptees had. This further indicated a strong genetic influence. In a Swedish study, 2,324 Swedish adoptees were studied and checked for a criminal record.[23] The study found that there were twice as many criminal sons of criminal fathers than criminal sons of non-criminal fathers, which again pointed to genetics. In a Danish study, 14,427 Danish adoptees were analysed.[24] They found that children who had a parent with a criminal record, and who were adopted into non-crime families, had a higher rate of being a criminal themselves. These independent studies allow us to conclude that criminal behaviour has a strong genetic component. The next question then is what the genes that give people a higher risk of being a criminal might be, and what might these genetic variants be doing to increase that risk?

The strongest evidence for a genetic link to criminal behaviour has been found in a gene encoding an enzyme called Monoamine Oxidase-A (MAO-A).[25] This enzyme is a key modulator of normal brain function. It controls the levels of three important neurotransmitters in the brain: dopamine, norepinephrine and serotonin (as we saw in previous chapters). These neurotransmitters have various functions, with dopamine being especially important for what is termed reward-motivated behaviour. The anticipation of most types of rewards increases dopamine levels in the brain and this makes us feel good. The main role of norepinephrine is to increase arousal and alertness. Serotonin in popular culture is seen as the happy neurotransmitter, though its functions are more complex, being involved in reward sensations, memory and learning. It's the job of MAO-A to keep this stew of neurotransmitters in check.

The first indication that something might be wrong with MAO-A in criminal behaviour was in a study of a large Dutch family who had a history of impulsive aggression and antisocial behaviour.[26] They carried a mutation

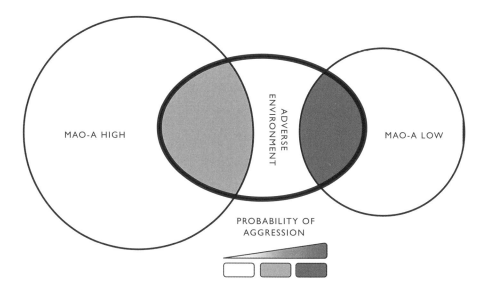

MAO-A HIGH

ADVERSE ENVIRONMENT

MAO-A LOW

PROBABILITY OF
AGGRESSION

in the MAO-A gene, which produced a form of MAO-A with lower activity than the normal form. Subsequent studies confirmed that having a less active form of MAO-A made people hypersensitive, and so were affected more by negative experiences and would act aggressively in defence. More importantly, those with the less active MAO-A who had also been abused as children were much more likely to commit crimes.[27] Other genes involved in

THE ACTIVITY OF A BRAIN ENZYME CALLED MONOAMINE OXIDASE-A HAS BEEN LINKED TO IMPULSIVE AGGRESSION AND ANTISOCIAL BEHAVIOUR. LOW LEVELS COMBINED WITH ADVERSE LIFE EVENTS IN CHILDHOOD IS A STRONG PREDICTOR OF THESE BEHAVIOURS. MIGHT CRIMINAL TENDENCIES BE WRITTEN IN OUR GENES?

controlling the same neurotransmitters as MAO-A have also been implicated in aggression and criminal behaviour, but MAO-A stands out because a large number of independent studies has supported its role with some truly excellent science performed in this area. The current challenge is to explain exactly why people with less active MAO-A have a propensity to be more aggressive and to commit more crimes.

Let's take a closer look at the evidence. The first thing that needed to be established to make the science robust was to actually define what is called the 'phenotype'. In this context, the phenotype is defined as 'the set of observable characteristics of an individual resulting from the interaction of its genotype [i.e. the genetic makeup] with the environment'. It means that the genetics of a person interacts with the environment that person is in to give rise to some characteristic. In this case, the characteristic that is used is aggression in response to an environmental trigger. It could be someone cutting into your lane if you're driving and you then respond aggressively. Aggressive behaviour is an important response to measure as it's easy to observe. It is also important to study, because of the devastating effects aggression has on society. For example, each year in the US, there are over five million incidents of non-fatal violent crime, costing over $200 billion, which include direct costs such as legal fees, medical expenses and incarceration costs. In Ireland in 2018 there were 19,995 non-fatal assaults.[28] Current treatments for pathological aggression are also limited, involving antidepressant medication and cognitive/behavioural therapy.

Studies into the aggressive phenotype start by classifying aggressive behaviour. There are two main types: proactive and reactive. Proactive is initiated by the offender and is directed towards a positive outcome for the aggressor, which could be dominance or theft. Psychologically, proactive aggression has been linked to what are called 'callous-unemotional traits', which include a lack of empathy and remorse. Reactive aggression involves an uncontrolled or exaggerated response to a perceived provocation or threat. People who react aggressively to a situation are said to exhibit 'hostile attribution bias', whereby they perceive provocation or threat to a situation that isn't especially threatening. It is common in people who have been maltreated as children.

As mentioned above, the first indication that something might be going on in the MAO-A gene was the Dutch family who had a mutation in the MAO-A gene. The men with this mutation were characterised by abnormal levels of disruptive, violent outbursts, often triggered by frustration, anger or fear. This in turn led to criminal behaviour, which in the family in question included attempted murder, rape and arson. Scientists noted other traits in the affected men, which included intellectual disability, sleep disturbance and a strange feature involving unusual hand movements. This finding was important for biological criminology and led to a resurgence in the field: but this particular mutation turned out to be extremely rare, and a second case was only found in 2014, 20 years after the first report.[29]

However, the spotlight was firmly on the MAO-A gene in criminal behaviour. The next important step came when scientists deleted the MAO-A gene in mice.[30] These mice were highly aggressive, similar to the humans who express low levels of MAO-A.[31] The scientists also observed higher levels of dopamine, norepinephrine and 5-HT in the brains of the mice. The levels of 5-HT reached a level ten times higher than in normal mice. The behaviour of the mice was interesting. In the first week of life they nodded their heads repeatedly and trembled. This was followed by hyperlocomotion and hyperreactivity and the onset of aggressive biting. The mice also exhibited exaggerated defensive responses to innocuous stimuli. Overall, studies in these mice strongly support the notion that the reason why the Dutch males were aggressive with criminal tendencies was in large part due to the defect in their MAO-A gene.

Other studies on the MAO-A gene have further strengthened its role as being involved in aggression. Numerous variants in the gene have been reported since the Dutch study, and the next best study reported on a version called uVNTR.[32] This version also leads to a lot less MAO-A being made. Several studies have shown an association with uVNTR and a tendency towards aggression, hostility and an antisocial personality. People with the variant are also less able to process facial expressions in others. Importantly, they also have a tendency towards reactive rather than proactive aggression.

In spite of the strong genetic link provided by studies into MAO-A, the environmental influence on people must also play a key role. The problem here is parents who carry the genetic variant that their sons inherit might be bad parents and this might affect behaviour. At a minimum, the environment is likely to interact with the genetics, which is termed 'nature via nurture', as we saw in the chapter on addiction. A son may be carrying the defective gene, and the subsequent behaviour of that son will be revealed by how they are brought up. A study in New Zealand has supported this, where male carriers of a defective MAO-A gene with a history of child abuse or neglect had a much higher incidence of antisocial behaviour than those who didn't have that history.[33] This was then confirmed in studies in the US, UK and Sweden. The New Zealand study was carried out over a 30-year period (psychologists need to be patient) and showed that children with the low-activity version of MAO-A who had been abused developed behavioural problems at the age of 16.

MACAQUE MONKEYS WHO HAVE LESS MAO-A AND WHO ARE RAISED WITHOUT THEIR MOTHERS ARE MORE AGGRESSIVE.

Studies in primates have also provided additional evidence that MAO-A levels are important for aggression.[34] There is variability in the MAO-A gene in many species, including macaques, gorillas, orangutans, chimpanzees and bonobo monkeys. Macaques in particular have been extensively studied, and those with a version that produces less MAO-A who were raised without their mothers exhibit much more competitive behaviour and aggression.[35] This provides striking confirmation of what is being reported in humans.

Curiously, studies have shown that a different version of MAO-A, which shows higher expression than normal, predicts risk of aggressive antisocial behaviour in females.[36] This is difficult to explain as based on the data in males, where lower expression of MAO-A links to aggressive behaviour, the expectation would be that females with higher levels should be a lot less aggressive. This illustrates the difficulty in these kinds of studies which are probably due to what are called modifiers: things that modify the effect of a given version of a gene. Testosterone could be a modifier here. Men have higher levels of testosterone and this may mean more aggression if the men have lower MAO-A. If there is less testosterone, as occurs in females, perhaps high levels of MAO-A become problematic. The gene for MAO-A is expressed on the X chromosome (of which females have two copies), which might also be important. Overall, this makes the study of the role of MAO-A in female aggression and criminal behaviour difficult to assess.

Given the strength of association between the MAO-A gene and the tendency towards aggressive behaviour leading to criminal activity, should genetic data be used in the courtroom to provide extenuating circumstances for criminal behaviour? A study in 2017 examined cases between 1995 and 2016 to identify how many court cases used low expression of MAO-A as mitigation for crimes committed.[37] It found nine US cases and two in Italy where MAO-A was used. Overall, two cases resulted in a lesser charge. In addition, five of the cases used MAO-A in appeals and two of these were granted sentence reductions. The authors concluded, however, that it was difficult to gauge the effect of the evidence on MAO-A on the outcomes.

One problem is that the strongest evidence for MAO-A genetics being important for aggression occurs when there is also an adverse childhood, which can be difficult to assess. There was a famous case in the US in 2009 when an argument was made that a combination of the person carrying the low expression version of MAO-A and a history of child abuse predisposed the person standing trail to commit murder.[38] As a result, the defendant in the case was spared the death penalty but was sentenced to 32 years in prison. Overall, however, there is no evidence that genetic evidence affects whether the judge or jury think the person being tried is culpable or not. They might conclude, though, that the defendant is guilty partly because of their genetic make-up, making them more likely to offend, and so they should be found guilty and jailed.

It has also been argued that people don't believe that genetics is a sufficient justification for why a crime is committed. Clearly if someone is deemed to have what is termed 'reduced behavioural control', either because they have a mental illness or because they are young, then there can be a justification for why a crime is committed, and that might in turn give rise to a different outcome – for example a custodial sentence in a mental health institution. This, however, is still a work in progress. If more research can be done on the mechanism whereby a defective gene impacts on behaviour and responsibility, as is happening with MAO-A, it may change how such genetic differences are viewed.

Although MAO-A has been the subject of the majority of studies examining the genetic basis for aggression and criminal behaviour, a host of other genes are being considered; these genes include those encoding proteins called TPH2, 5-HTT, the D4 receptor and COMT.[39] Similar to MAO-A they all impact on the neurotransmitters serotonin, dopamine and norepinephrine, further emphasising the potential importance of these neurotransmitters in criminal behaviour.

Given all of the above, I am still left with the question – why it is that I wasn't an inmate in Mountjoy listening to someone give a lecture? The answer probably lies in the fact that I had a loving and stable upbringing, didn't suffer any hardships as a child, had a peer group that didn't have any criminal tendencies (probably thanks to my mother, who made sure I went to a good school and always kept an eye on who my friends were) and lastly, in all likelihood had a genetic makeup that didn't predispose me to becoming a criminal, given more adverse circumstances. In other words, I was lucky. What can we do to help others who are not so lucky, either to decrease the chances of them becoming criminals or to try and be sure that they won't re-offend? We have to support those in need of help and to try to reduce the emotional pain that might lead to criminal behaviour. It has to start with our children.

In 2015 experts were asked what we need to do to decrease the crime rate in the world's most violent cities, which are mainly in Latin America and the developing world.[40] Their suggestions, which could be applied anywhere in the world, make for compelling reading. We need to treat violent behaviour as a public health concern, using technology to reach every child so that they all feel cared for through parenting interventions and early childhood education. Family and community support are essential. Overly repressive and punitive policies don't work, and governments must go beyond law enforcement and criminal justice. Proactive community and school programmes must be supported and encouraged. Worryingly, having fallen for some years, crime rates appear to be on the increase again in Ireland and it's not clear why that is.[41] But it must take our closest attention. **The bottom line: no one – victim, perpetrator or society as a whole – wins when it comes to crime, so we need to do more to understand why it happens and try to prevent it from happening in the first place.**

CHAPTER 9

WHY DO YOU STILL THINK THAT MEN ARE FROM MARS AND WOMEN ARE FROM VENUS?

—

'Behind every great man
is a woman rolling her eyes.'

—

Jim Carrey

W HEN I WAS a teenager my sister Helen encouraged me to join Amnesty International. When my membership card arrived, my mother confiscated it, fearful that I would be somehow led astray by a radical organisation. This made my sister laugh, annoying my mother even more. We joined forces on that one. Amnesty was just one organisation that Helen was, and continues to be, involved in. She is my hero because she has spent her life working with vulnerable people, the people who most of society wants nothing to do with, helping them in all kinds of ways. Being a social worker means she was always a radical lefty type who ate a lot of mung beans and stank out my mother's kitchen with exotic smells, my mother being of that generation of Irish people who insisted on potatoes with everything.

Helen went to the Philippines in the 1970s to work as a teacher – unusually for a young woman at the time – of young Filipino girls in seriously deprived areas. Liberation through education. Women's rights, especially in Ireland, has always been her key issue. In 1971, at the age of 15, she was on the 'Contraceptive Train' that brought condoms from Northern Ireland to the south (where they were illegal until 1985). After the Philippines, she went to London to study social work at Middlesex Polytechnic. (Unlike me, her privileged male sibling who could go to university, she was a Polly Wally – as we called polytechnic students – although it's now called Middlesex University). She lived in a flat on the 13th floor of a tower block in London's East End. She turned it into a refuge for Irish women who had gone to London for an abortion (which was illegal in Ireland until 2018). She has always fought for decency, for what is right and, most of all, for women.

Men versus women: this distinction is fundamental to us humans, and the differences between us are the cause of much of the joys and troubles in our lives. At one time it seemed so simple. Based on observation and measurement, men had one X and one Y chromosome, and women had two X chromosomes. A man was physically stronger and had a deeper voice, a penis and testicles, more body hair and fewer curves. A woman was physically less strong and had a higher voice, breasts, ovaries, a uterus and a vagina, had periods and could have babies. Because of prejudice and ignorance, there was also a time when men were deemed to be more logical and intelligent, more likely to be leaders, much better drivers and better at maths. Women, on the other hand, were deemed the weaker sex, who needed doors to be opened for them, weren't worldly enough to have their own bank account and couldn't possibly drink a pint of beer. (My sister always drank pints, much to my father's annoyance.)

You mean a <u>woman</u> can open it ?

Easily—without a knife blade, a bottle opener, or even a husband! All it takes is a dainty grasp, an easy, two-finger twist—and the catsup is ready to pour.

We call this safe-sealing bottle cap the Alcoa HyTop. It is made of pure, food-loving Alcoa Aluminum. It spins off—and back on again—without muscle power because an exclusive Alcoa process tailors it to each bottle's threads

after it is on the bottle. By vacuum sealing both top and sides, the HyTop gives purity a double guard.

You'll recognize the attractive, tractable HyTop when you see it on your grocer's shelf. It's long, it's white, it's grooved—and it's on the most famous and flavorful brands. Put the bottle that wears it in your basket . . . save fumbling, fuming and fingers at opening time with the most cooperative cap in the world—the Alcoa HyTop Closure.

Alcoa Aluminum

THIS 1953 ADVERT DISPLAYS A MAJOR TECHNOLOGICAL ADVANCE: THE BOTTLE CAN BE OPENED EASILY 'WITHOUT A HUSBAND!'

MALE AND FEMALE MANDRILLS ARE WHAT
ZOOLOGISTS CALL THE 'MOST SEXUALLY
DIMORPHIC' OF MAMMALS. THE MALE WEARS A
LOT OF MAKE-UP.

Nowadays, all this has changed, changed utterly, with all the terrible beauties that have been born. Gender has become fluid – a person's sex may be male or female or, in rare cases, intersex[1] based on their biology, but their gender might be different from their sex. Equally, it turns out that with each study done, the differences between men and women, apart from the physical ones, become less and less clear. Or even switch entirely. What is the truth of the matter? Are men really from Mars and women from Venus? Apart from the physical ones, which are easy to describe, what are the scientifically proven differences between men and women and what do these differences mean for society? The questions are fraught. The studies that have been done often exemplify the biases scientists bring to their analyses. 'What?' I hear you say. 'Scientists can be biased?' I'm afraid they can. But, as with all of the issues in this book, we just need to (wo)man up and do our best to reach conclusions based on the science. As with all science, we need to ask what the most reliable data tells us. There are differences, but not the ones you think. One memorable study that turned out to be biased, although not related to the differences between men and women, concerned menstruation. I bet you believe that women's periods synchronise when they spend time together? Well, this has been proven not to be the case.[2] Get ready for more surprises.

In the animal kingdom, things are more straightforward. Some species show marked differences between the sexes. One of the best examples of this is the mandrill monkey, which is the most sexually dimorphic mammalian species. The males wear a lot of make-up on their faces and behinds. (Well, it's not quite make-up, as the vivid colours are natural.) There is also a major difference in size between male and female mandrills. Males are three times heavier than females, bringing the risk of the male flattening the female during sex. Mandrills are a bit like pheasants: the male pheasant has exotically coloured feathers, a large flamboyant tail and a long appendage around its eyes called a wattle, while the female is small and dull. But mandrills and pheasants are nothing when compared to the delightfully named triplewart

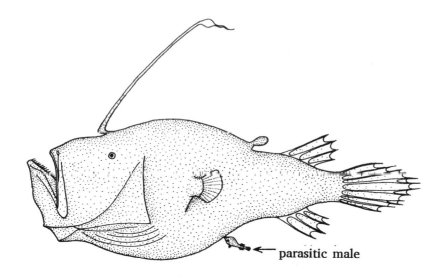

parasitic male

seadevil. These creatures live 2000 metres underwater. Females are 300 cm long, while males only reach 1 cm in size – I suppose it makes sharing a bed easier with no competition for the duvet. Examining our primate cousins also reveals differences between the sexes, although these lessen as we get to the species that are closest to us. As they reach sexual maturity, male orangutans begin to develop enlarged cheek flaps that are meant to exhibit their dominance. When there is more than one male within a family, the more dominant male will exhibit more exaggerated cheek flaps. Who would have thought dominance would come down to jowly cheeks? But when we get to chimpanzees and bonobos (our closest relatives) the males and females physically differ in the same way that we differ: slight differences in size and muscle strength and clearly differentiated sexual organs.

So what are the differences between men and women, apart from physical ones that are plain to see, that scientists agree on? A good way to examine traits in humans is to use a bell curve (because it's shaped like a bell), also known as a normal distribution. If we measure the height of lots of men and women and then plot a graph of height versus the frequency of each height (meaning how common each specific height is) in the population, you get a curve. A small number of people will be small and a small number will be tall, and the rest will fall in the middle. When we measure height in women and men, we see different curves for each sex.[3] This tells us that men on average are taller than women, although there will be a lot of women who are taller than a lot of men. If we look at the bell curves for aggression, again we see that on average men are more aggressive than women.[4] This might be due to the hormone testosterone, which is also why men have more muscle mass than women. But as with all the traits we are discussing here, there will be overlap: some women are more aggressive than some men.

Where else do we see clear differences? One fascinating area is differing susceptibilities to diseases.[5,6,7] Ninety per cent of cases of primary biliary cirrhosis (an autoimmune disease of the liver) are in women, while primary sclerosing cholangitis (an inflammatory disease of the bile ducts) is much more common in men. Many autoimmune diseases are more common in women (e.g. Sjogren's syndrome, multiple sclerosis, scleroderma and lupus). The bone-wasting disease osteoporosis is four times more common in women. COVID-19 is equally common in men and women, but men fare worse, with 70% of those who die being men. And as of April 2020, men have died at 1.4 times the rate of women.[8] Women are ten times more likely than men to develop eating disorders, such as anorexia nervosa and bulimia, and are twice as likely to suffer from depression. Men, on the other hand, are four times more likely than women to have autism and 1.4 times more likely to suffer from schizophrenia. They are also twice as likely to die by suicide. The difference is especially marked in Ireland, where four out of five suicides in 2018 were male.[9]

Why these differences in disease susceptibility exist is mostly unknown. Osteoporosis is an exception since it is known that the hormones oestrogen and progesterone keep bones strong and the level of these hormones falls after the menopause, which is when osteoporosis happens. For other diseases, though, the question of differing disease susceptibility in women has, rather scandalously, largely been ignored by medical researchers, although that is starting to change. As is the case with osteoporosis, scientists suspect it's to do with differing levels of hormones, but that has yet to be clearly proven, especially when we look for precise mechanisms. A recent study examined sex differences in the expression (meaning how much a gene is turned on) of 12,000 genes in 12 different tissues from human, monkey, mouse, rat and dog; it's a bit like determining the volume level on these genes and how far the knob for each gene is turned up. Crucially, differences were found between males and females,[10] who seem to use the same genes in different ways. The differences explained 12 per cent of the variant in height between male and female (where males on average are 13 centimetres taller than females). Hundreds of genes associated with height had different levels of expression between males and females. The work is causing much excitement in medical circles, as further analysis might reveal why men and women differ in susceptibility to different diseases. If the volume control on genes that might explain these differences could be turned down, it might lead to new treatments.

Even though more women suffer from diseases, they still live longer on average than men. In Ireland life expectancy for women currently stands at 84 years, while for men it's 80.4 years.[11] Again, the reasons for this difference are not fully understood.[12] Longevity in women might seem like an advantage, but women also have higher rates overall of physical illness, more disability days, more doctor visits and more hospital stays.[13] Several reasons for longevity differences have been proposed. Men tend to have more fat surrounding their organs (called visceral fat – see Chapter 4), whereas women have more fat under their skin. This is determined in part by oestrogen, and it matters because fat surrounding organs predicts heart disease – a major cause of death in men. Lifestyle differences may also be important, with men being heavier smokers than women for much of the twentieth century, although that difference has lessened and it may be one reason why sex differences in mortality are likely to narrow in the future.[14] It will be intriguing to see whether life expectancy between men and women becomes more equal in the future.

Once we move away from the physical and medical areas, and towards the mental, things get a lot murkier. Often the differences reported are small – possibly too small to be significant – yet much is made of them. We will proceed with caution.

The current consensus is that men and women perform equally well on IQ tests.[15] In one major study, males and females were found to be highly similar when it came to skills such as reading comprehension, mathematics, communication skills and motor skills (such as manual dexterity). From a review of 46 studies, 78 per cent of sex differences in these kinds of skills were small or close to zero and overall, bigger differences are evident between men (and between women) than between men and women.[16] This is important, since any differences in these kinds of traits may not be explained based on sex differences alone.

Across 4,000 separate studies (which gives you an idea of how exhaustive and exhausting these kinds of studies are), the gap in achievement at maths was largely zero, putting to bed once and for all the anecdotal and discriminatory notion that there is a sex difference in mathematical skills.[17] Then there was the daddy – or mammy – of all studies comparing men and women, which bravely addressed personality differences.[18] Thirty-one thousand personality tests were analysed (again, a lot in these kinds of studies) and the researchers examined personality traits such as warmth, emotional stability, assertiveness, gregariousness, dutifulness, friendliness, distrust, imagination, openness to change, introversion, orderliness and emotionality. They were trying to address whether these traits were comparable between men and women. And guess what? They found that women scored higher on anxiety and warmth, while men scored higher on dutifulness and assertiveness. The other traits showed more minor differences. The robustness of the study can be seen in the observation that 95 per cent of the time, the personality profile of a randomly picked male will be more male-typical than that of a randomly picked female. Knowing the personality profile of an individual makes it possible to correctly guess his or her sex about 85 per cent of the time, which is pretty powerful. Of course, the precise reason why women show more warmth but are inclined to have higher levels of anxiety while men are more dutiful and assertive remains unknown, and again is likely to be down to a combination of upbringing and a hardwired difference, although the relative proportions of these are difficult to figure out.

One area where women supposedly shine is empathy, which is defined as the ability to read other peoples' thoughts and feelings. Surely, I hear you say (what with all your baggage and preconceived notions), women will outperform men in that area? Some studies do show that, but if the test is not labelled as an empathy test, any difference goes away, suggesting that it is the tester who shows bias:[19] if they know empathy is being tested, they will grade women higher than men. And remember, these tests often come down to filling in questionnaires, which can suffer from being subjective.

Another problem with these kinds of studies is how we define warmth or empathy; yet several studies have concluded that women have more emotional intelligence than men.[20] This is defined as the capacity to express one's emotions and to handle interpersonal relationships judiciously and empathically. The assumption is that women are better in this capacity than men. Women have been shown to respond differently to someone who is upset or emotional; they are more likely to become upset themselves when they see someone else upset and these feelings will persist.[21] A part of the brain called the insula is involved in this response. Men, on the other hand, will sense the feelings for a moment and then tune out of the emotion. This tuning out might lead to other parts of the male brain being activated, which are more to do with problem-solving, although this hasn't been proven. Men tuning out emotionally is a common complaint of women. Both responses actually have advantages. If it's a heterosexual couple, perhaps it's advantageous if one can protect themselves from distress and try and find a solution that's urgently needed. The other provides support and nurture to the person who is distressed. Males and females could, of course, perform either role, but arguably the view is the female is inclined to be the supportive one.

But when psychologists look at leaders in companies, key traits transcend sex. Successful leaders of both sexes show empathy that will persist and then become problem-solving – a mix of both traits.[22] Is this because the women in leadership roles have learnt to become more like men, or did they have these traits from the start, which led to them being promoted to leadership roles? We don't know. Scientists have made a similar observation in chimps. When a chimp is distressed female chimps are more inclined than males to give solace to the troubled chimp, perhaps by stroking to calm it. However, the alpha male in the chimp troupe gives solace more often than any female.[23]

Leaders, be they male or female, need to be empathetic.

The issue of women and leadership in the workplace is an increasingly important one. Until recently (with notable exceptions such as Angela Merkel, Margaret Thatcher, Jacinda Ardern and Indira Gandhi) most leadership positions have been held by men. The increased prevalence of male bosses over female bosses appears to continue unabated. Men are more likely overall to be made the boss, possibly because of assertive personalities. In 2018 only 24 Fortune 500 companies were led by women, even though women have the same level of professional qualifications as men.[24] So why are women not being made leaders?

One reason might be expectations. In

ANGELA MERKEL BECAME
CHANCELLOR OF GERMANY IN 2005.

a study in 2015, 60 per cent of male employees expected their employers to play an active role in their career progression, whereas only 49 per cent of women had this expectation.[25] Men are more assertive when it comes to promotion, with women being more hesitant about speaking up about their career ambitions. Another reason might be their male managers behaving badly. In a recent study, more than half of women in middle and frontline management roles have had a manager take credit for their work. This might enhance the manager's chances of promotion, which is then denied to the woman who actually did the work. Another study has shown that men tend to be more career-centric, wanting to maximise their financial return from work.[26] Women might view work more holistically, approaching their careers in a more self-reflective way, valuing meaning, purpose and connections with co-workers,[27] although again this needs to be examined scientifically.

These traits, though, might actually mean that women make better leaders. In a recent survey, a number of traits were emphasised by women in leadership roles[28] – again, this is anecdotal and more evidence is needed. Yet women leaders feel that they are more inclined to nurture team members, which helps the team develop their own skills. Women might also focus more on teamwork overall and are less egotistical, which is seen as a good thing as ego can sometimes get in the way of a good decision. Women are also more inclined to be able to balance multiple challenges both in work and in their personal lives (which might include managing the household and ageing parents, tasks that still more commonly fall to women). They may not be innately better at multitasking, rather it's that they have to do it more often and become good at it. They bring that skill to their role as a leader. In spite of all these skills, one study has shown that when asked about preference for a male or female boss, an interesting result emerged:[29] women prefer a male boss 39 per cent of the time. It was even lower for males: only 26 per cent of men said they would prefer a female boss. When it comes to leadership, gender stereotypes might still prevail, at least for men, who perhaps see themselves as being emasculated by a female boss. Yet again, we are longing to see a bell curve here. There will be a mix of abilities in both males and females and perhaps the curve for the female boss will be more to the right than the curve for the male boss.

The kinds of analyses that lead to conclusions on emotional and personality differences between men and women usually involve the filling in of some kind of questionnaire. They also involve people in the surveys identifying as male or female, which may not be consistent with their gender. Surveys can, of course, be useful, but another, probably more robust way to determine difference is to examine male and female brains. A lot of nonsense has been published in this area. In 2005 there was a study of the brains of 21 men and 27 women (the small numbers immediately make us suspicious).[30] A major difference was purported to be found between their brains – and the media went into a frenzy. It was claimed that men had 6.5 times as much grey matter, whereas women had ten times as much white matter in their brains. Grey matter consists of neuronal cell bodies. White matter, on the other hand, has a lot of what are called myelinated axons. These are vast differences, and the media crowed about how this explains why men are better at maths (which might involve the grey matter) while women are better at multitasking (which might involve the white matter). In truth, it hasn't been proven that proficiency at maths or multitasking is due to grey or white matter. As was recently concluded in the world's leading science journal, *Nature*, 'The history of sex-difference research is rife with innumeracy, misinterpretation, publication bias, weak statistical power, inadequate controls and worse.'[31]

Psychologist Gina Rippon has likened the field to 'whack-a-mole'. A study is published claiming a difference between men and women. This is used to taunt political correctness until other researchers reveal a flaw in the study. The mole is whacked, but then another one pops up. The examination of the brains of men and women is full of such mole-whacking, and, in spite of many efforts, no convincing physical difference between the male and female brain has been found. If you give a neuroanatomist a brain and ask him or her to say what sex it is, they won't be able to. Although, having said that, there is a region in a part of the brain called the medial preoptic area, known as the sexually dimorphic area, which is larger in males than in females in at least nine different species, including in humans.[32] Its size fluctuates in both sexes, though with an overall shrinking in both sexes as we age. Its size has been linked to libido and sexual orientation, although this link remains controversial. In addition, a study in 2017[33] of 2,750 women and 2,466 men found that women have thicker cortices (the convoluted surface layer of grey matter in the brain) than men, while men have slightly higher brain volumes. Again, however, differences between men were greater in these features than between men and women, which agrees with studies showing greater variation in IQ scores between men than between men and women. Overall, it was concluded that you would be hard-pressed to say whether a brain had come from a man or a woman.

Female brains are smaller, but this is in proportion to overall body size. In the nineteenth century, when this was noticed, it was known as the 'missing five ounces'. Perhaps it was in those five ounces that we would find the holy grail of the difference between men and women – but they were only missing because the overall size of a woman's body is smaller. There was also a now-infamous study in 2014, which claimed that the connections in a woman's brain were *between* the brain's hemispheres, whereas the connections in the male brain were more *within* the one hemisphere.[34] Yet most connections weren't even mapped, and there was no control for puberty-related maturation or, yet again, overall brain size. How the study was accepted for publication remains a mystery. Clearly, the neuroscientists who accepted the paper need their heads examined ...

Given that few physical differences between the male and female brain can be convincingly shown, the current view is that a gendered world will produce a gendered brain. The differences in personality and emotional intelligence are now believed to be primarily due to society as opposed to being innate. Most psychologists now agree that any differences are a result of the pink-versus-blue culture that the brains of babies and children are soaked in from the moment a parent knows the sex of the foetus. How do we know this culture exists? A study of 124 primetime television programmes in the US showed that women have mainly interpersonal roles – romance, family and friends – whereas men have work-related roles.[35] A study of 5,618 children's books revealed that males are represented twice as often as females in book titles and 1.6 times as often as central characters.[36] These kinds of influences, and many more like them, might go a long way towards explaining why women and men turn out the way they do. But we mustn't forget the study mentioned earlier, in which differences were observed in the 'volume control' of male and female genes. There may be no physical difference (say in connections between parts of the brain) or differences in actual genes, but there may be differences in the level of expression of the same genes between men and women. This might be controlled by sex chromosomes or environment. More work on that aspect in the brain could indeed reveal the basis for differences in psychological traits. We can eagerly anticipate further studies in this area.

The differences between men and women, whatever their basis might be, become significant when we consider two things: education and the workplace. First, let's look at education. The sharp end of the male-ver-sus-female debate is performance in school. Study after study has confirmed that boys perform worse than girls in school, with those coming from low-income backgrounds doing especially badly.[37] So here we have a clear verifiable difference. Why is this, and what can we do about it?

In developed countries, boys are on average much worse at reading and less likely to go to university. They are still slightly better at maths in school, but the gap is shrinking and in some countries (Scandinavia and China) it's gone.[38] Ireland is the same – girls outperform boys at Junior Cert and Leaving Cert in most subjects (but still not maths).[39] Growing numbers of adult men still live with their parents: in the UK a third of men aged between 20 and 34 do so, whereas for women it's a fifth. Jobs dominated by poorly educated men (such as manual labour or driving) are the most likely to become automated.

Discontented, less-educated men vote for right-wing populists like Donald Trump, so if we don't do something about this education gap we will reap the whirlwind of divisive politics that is often damaging for us and planet Earth. Surprising as it may seem, boys' grades were once deliberately upgraded (for example, in the eleven-plus exam in the UK).[40] In Japan, universities have admitted to discriminating against female applicants by manipulating exam scores. Girls were also more likely to be ignored by teachers or not given access to important subjects. Ever since this kind of sexism diminished, girls began outperforming boys.[41]

Where men once went to college in proportions far higher than women – 58 per cent to 42 per cent in the 1970s – the ratio has now almost exactly reversed.[42] The reason is that more women apply to college. There's evidence of boys losing motivation from the age of eight. The vocabulary gap at age 11 is wide. The literacy gap peaks at 16. In the programme for international student assessment (PISA) scoring system used by the OECD, the lowest-performing pupils are male. So why is all this happening? Are girls just better than boys in school?

Girls mature earlier – by as much as two years; their brains have quantifiably more grey matter (the thinking part of the brain) at a younger age. It's possible that they could be given more attention by teachers and so learn more. Co-educational schools are seen as more female-friendly, with most classrooms now female-led. Ninety-seven per cent of early childhood educators are female. Female teachers dominate at primary and secondary level where they outnumber males two to one. Female educators are, however, under-represented in universities, but only at the professorial level.

Ironically, the gender gap in educational attainment (for example in exam performance) is worse in the most gender-equal of countries. It's not fully clear why this is. The big challenge now is how to make classrooms boy-friendly without disadvantaging girls. One answer might be a return to single-sex schools; yet girls in single-sex schools have higher levels of exam stress compared to boys.[43] In addition, girls in single-sex schools tended to be more negative about their experience of school than boys. There is more pressure to perform socially for girls in single-sex schools. Girls do better academically in single-sex schools, but the pressure can be too intense for many and so they ultimately end up underperforming in all spheres. The psychologist Oliver James has recently identified high-performing 15-year-old girls as the unhappiest group of people in England or Ireland.[44]

On average, and all things being apparently equal, girls generally outperform boys in the classroom, although comparing bell-shaped curves would be useful to quantify the magnitude of the difference. A study carried out by Professor Emer Smyth of the ESRI found little consensus on whether sex segregation leads to better or worse academic outcomes.[45] Once boys reach the age of 18, the gap narrows, partly because weaker male students have left education but also because boys' brains catch up with those of girls. The main reason for the difference in performance between boys and girls in education might be the level of maturity at different ages. So perhaps boys should start school a year or two later?

What else might be done? How boys up to age 16 are being educated is being examined to make it more boy-friendly. Boys might like more structure and rules (although that needs to be proven) and badly behaving boys need support, not punishment. Lessons in vocabulary help boys more, and boys should be encouraged to read whatever they want. All of this helps. As does training teachers about their own gender bias. Teachers are being encouraged to value risk-taking behaviours in boys instead of suppressing them. We also need to encourage more men to enter teaching: Germany is leading the way with an active recruitment programme.[46] Generally, though, the problem has received little attention as elite boys always do well and men overall still get paid more and are promoted more in the workplace. But as women catch up and as this particular gender gap continues to grow, there will be a crisis: lots of poorly educated men in low-paying jobs, leading to a worsening of a whole range of societal problems, including ending up in jail, as we saw in Chapter 8.

When it comes to occupations, big differences are evident between what men and women do. Men are more likely to work with things, while women are more likely to work with people. Again, this could be a result of the pink-versus-blue brainwashing during childhood. Why are there so many more male engineers than female? Parental and societal influences are likely to play a role, and one study has shown that women are often discouraged from working with computers. Since the 1980s the proportion of female graduates in science and medicine has gone up, while the number of female computer scientists has gone down.[47] When it comes to medicine, interesting patterns emerge:[48] just under half of all doctors across developed countries are female. In Japan and Korea only 20 per cent of doctors are female, while in Latvia it's 70 per cent. In Ireland 41 per cent of all doctors are female, with 42 per cent of GPs being female and 39 per cent of consultants. Another hot topic in the workplace that has yet to be fully resolved is the gender pay gap. Studies carried out by the European Commission reveal that women earn 16 per cent less than men per hour for the same job:[49] this is illegal under the Employment Equality Act. The situation is worse for lower-paid jobs such as cleaners or street vendors, where the gap can be as high as 39 per cent. Yet the signs are hopeful. Analysis of the gender pay gap by age reveals that for women under 40, the gap narrows to 2.6 per cent. But in Hollywood, for those of you who are movie stars, the situation is especially grim for women. A recent study has shown that female actors earn substantially less than male actors with similar experience. A study examined 1,344 films featuring 267 stars and on average the pay for women was $1.1 million less per movie, for no good reason other than the movie bosses could get away with it.[50]

In spite of all this, a girl in a western country can expect to be able to do whatever she wants in terms of career, and indeed any other activity. This represents tremendous progress compared to the lot of girls until relatively recently. Yet issues remain around, for example, childcare and the sharing of the domestic load. How will we know when equality is truly here? Christmas is a case in point. Studies have shown that women in the UK still do the heavy lifting when it comes to Christmas: sending Christmas cards, buying presents, doing the shopping and cooking the Christmas dinner.[51] When these tasks are shared 50:50, we'll be closer to equality. As the famous statistician Hans Rosling has noted, the advent of the washing machine did more for women's liberation than any other single thing. Another sign of equality will be the number of women owning toolboxes, which remains an almost exclusively male domain, although not in my house where my wife owns the toolbox. This shouldn't be the case, given the lack of difference in competency when it comes to fixing things (I'm not including myself in that). Once all women have toolboxes, we will know that they are finally free.

We also have to continue to correct misapprehensions. The psychologist who led the analysis of the 46 studies on sex differences (which didn't reveal any major differences between the sexes) mentioned earlier in this chapter has said that if we assume women are more empathic and emotional than men, and that men are more assertive, this can lead to biases in the workplace, which hold women back from promotion. It can also dissuade couples from trying to resolve conflict. Importantly, if there are differences, these should be treated more like deficiencies that can be helped. Society can change.

One fascinating study in India showed that the introduction of cable TV led to a significant decrease in reported acceptability of domestic violence against women and an increase in female autonomy.[52] This was because many female characters in popular soap operas have more education, marry later, have smaller families, work outside the home and hold positions of authority. Cable TV has also been linked to mothers having a decreased preference for giving birth to a son over a daughter in a culture that has commonly favoured sons.

And what about in Europe? French toymakers recently signed a pact to rid games and toys of gender stereotypes.[53] A charter was signed between toymakers and the French government, which aims for 'balanced represen-tation of genders in toys'. The French junior economy minister, Agnès Pannier-Runacher, has said that many toys project an 'insidious' message that discourages girls from becoming engineers or computer coders. She has

blamed toymakers of making toys for girls that are mainly about domestic life, whereas those for boys are more about construction, space travel and science. (The staff at France's major research centre, the Centre National de la Recherche Scientifique (CNRS), is 38 per cent female, yet only 10 per cent of these women are coders.) Gender neutrality in toys might change these sorts of percentages.

MANY DIFFERENT TYPES OF BARBIES ARE NOW AVAILABLE – EVEN SCIENCE BARBIE.

A recent study on gender stereotyping in Ireland has also revealed something interesting. When 5–7-year-olds were asked to draw an engineer, 96 per cent of the boys drew a male engineer, while just over 50 per cent of girls drew a female engineer.[54] The study was carried out by students Cormac Harris and Alan O'Sullivan, and the project was deemed overall winner at the highly competitive 2020 Young Scientists Exhibition. The work also listed resources that primary-school teachers can use to combat gender stereo-

MEDB, QUEEN OF CONNACHT
IN THE ULSTER CYCLE OF IRISH
MYTHOLOGY. SHE HAD SEVERAL
HUSBANDS AND WAS DESCRIBED
AS STRONG-WILLED, AMBITIOUS
AND CUNNING – THE ARCHETYPAL
WARRIOR QUEEN.

typing among young children.

Another long-running study analysed what schoolchildren draw when they are asked to depict a scientist. Over five decades, 20,860 drawings by students between the ages of five and eight were analysed (yep, that's a lot of drawings). One has to wonder where all these drawings were kept. In the 1960s and 1970s, fewer than 1 per cent of the pictures depicted scientists as female; by 2016, that percentage was 34. It was even higher, at 50 per cent, if the participant was female. This parallels the increase in actual numbers of female scientists in the US, which between 1960 and 2013 rose from 28 per cent to 49 per cent in biology, 8 per cent to 35 per cent in chemistry and 3 per cent to 11 per cent in physics.[55] Things are definitely getting better in the science world, even, to some extent, in physics.

Perhaps the Irish might serve as a model for the future of male/female equality. The Irish for 'men' is *fir*, and 'women' is *mná*. An English friend of mine thought it strange that in a pub I took him to the toilet labelled 'F' was for males and the one labelled 'M' was for females. He found this out when he walked in on a woman – and was told where to go. He wondered if this reverse nomenclature was an example of perverse Irish humour.

Celtic women were distinct in the ancient world, having many rights that other societies didn't have. I suspect my sister is the reincarnation of a powerful Celtic woman. Ancient Celtic women served as both warriors and rulers. Cuchulainn, Ireland's greatest mythological warrior, was trained by Scáthach, a female warrior from Scotland. Scáthach and her sister Aoife both led armies in battle. Aoife was also Cuchulainn's lover (he obviously met her through her sister, not an uncommon event in Ireland to this very day). Women often joined men in battle, using psychological techniques including loud screeching and wild dancing to scare the enemy. This behaviour can still be seen in nightclubs in Dublin, where Irish women continue to terrify Irish men. Boudicca and Cartimandua were both famous leaders of the Celts in Britain, as was Queen Medb in Ireland. Women were not systematically excluded from any role in ancient Ireland. They could become druids or diplomats and conduct business without their husband's consent. Marriage was viewed as a partnership. Each brought an equal dowry and women could own property independently of their husbands. Divorce was a simple matter as marriage was viewed as a contract. There is even evidence of polyandry – women having more than one lover or husband.[56]

The Irish throughout history have had a particular opinion on the relationship between men and women, as many proverbs show. How about this one: 'A woman is a woman when her man is a man.' So maybe the Ancient Irish got it right. **The bottom line: when it comes to men and women, it's our collaboration, complementarity and independence that counts.**

CHAPTER 10

WHY DO OTHERS SCARE YOU?

—

'She gave a startling interpretation, which
jolted the audience out of its complacency.
This was exactly what I wanted the song
to do and why I wrote it.'

—

Abel Meeropol, describing the first performance by
Billie Holiday of his anti-racist song 'Strange Fruit'

N 1985 I MOVED TO LONDON. At that time, the IRA bombing campaign was happening. I went for a drink with my father the night before I got the ferry from Dún Laoghaire to Holyhead (a standard route for Irish emigrants in those days). Although my father was born in Ireland, he had grown up in Salford, England, and considered himself English. He said to me, 'You might think you're smart, but you'll always be a Mick over there.' Jokes about Irish people being stupid or criminal were commonplace in England at the time. This was a mild form of racism, and I would laugh the jokes off, but they always hurt. And of course, being a white scientist in Ireland, I haven't experienced this feeling since. But why are so many of us racist, and what can we do about it? We live in a time when this question is more relevant than ever.

Around 180,000 years ago, a tribe of possibly no more than a few hundred people left Africa and entered the Middle East.[1] They stayed there a while, and their descendants went travelling. Seventy thousand years ago, some of them went to Asia. They reached Australia 50,000 years ago. Fifteen thousand years ago, some of the people in Asia eventually made it to North America and went on to populate that entire continent. Forty thousand years ago, those left behind in the Middle East moved north to Europe, settling all parts by around 10,000 years ago. All of the descendants of those people who left Africa 180,000 years ago eventually populated almost every corner of the earth

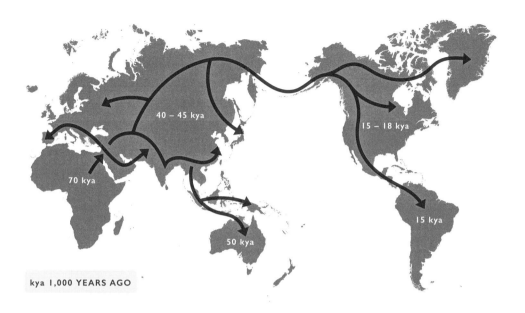

40 – 45 kya

15 – 18 kya

70 kya

50 kya

15 kya

kya 1,000 YEARS AGO

by using the defining feature of our species, ingenuity, to cope with local conditions: wearing animal fur in colder climes, using fire, or building boats to sail the Pacific.

HUMANS EVOLVED IN AFRICA AND GRADUALLY MOVED AROUND THE WORLD.

When different tribes that emerged over thousands of years in the various regions encountered one another, it was like distant cousins meeting up. They often fought, most likely out of fear that a rival tribe would take all their resources or abduct women for procreation. Or maybe we just like fighting.[2] Some anthropologists are of the view that we have an innate tendency towards violence, with one, Richard Wrangham, describing us as 'the dazed survivors of a continuous five-million-year habit of lethal aggression'.

In Europe technological advances meant the building of ships and navigation techniques that allowed the family of humans to reunite, first in the Americas when Christopher Columbus encountered Native Americans, and later in Australia when Captain Cook encountered native Australians. In both cases this led to genocide, with the invaders killing the locals, either actively or by spreading diseases like smallpox. Meanwhile, all over the world, tribes fought. They were even fighting in Africa before humans moved into the Middle East. Europe was almost constantly at war from the Dark Ages on, with the fighting culminating in two wars, World War I and II, which would leave whole generations of young men dead. Estimates of numbers killed range from 15 to 19 million in World War I, and 70 to 85 million in World War II. In Germany, one tribe (so-called Aryan Germans) tried to wipe out another – killing 6 million of the Jewish people. In Japan, the Americans managed to drop two nuclear weapons, killing up to 226,000.

What could have possibly justified all that killing? After all, these warring tribes are all descended from the first tribe that left Africa. Why do we hate each other so? Why do others frighten us? Is it an instinctive response to protect our resources when someone we don't recognise as part of our family invades our territory? And what can we do to stop all the fighting, which still goes on today, both in conflict around the world in racism and discrimination that still holds people back? Is there such a thing as different races at all? (Answer: a resounding no.) Why can't we listen to people we've made into heroes, like Gandhi or Martin Luther King, and stop all the fighting and trouble and just get along?

The instinct to be suspicious of others would appear to be hardwired. It occurs in all species (at least as far as we know), and, as with all conserved biology, it has been retained due to the survival benefits that it provides. Our first responsibility is to our own family. We will defend them against others should they be threatened, because we want our DNA to be passed on – a key driver for all life on earth. Or should I say that our DNA compels us to pass it on to the next generation? Early humans would have lived in small groups of maybe a hundred or so. You knew everyone and you were related to several of them, so it was also in your interest to defend your tribe. Many will also have some of your DNA and a larger group would provide more protection. Anthropologists who study so-called 'traditional' people have examined how different tribes interact with one another. Violence will happen if resources are threatened. The Hadza people in Tanzania have been studied extensively and give us

a good example of what life was like for tribes since *Homo sapiens* evolved.[3]

In 2015 there were around 1,200 Hadza living around Lake Eyasi in the central Rift Valley, with about 300 surviving by traditional means, exclusively living on foraging for food. Socially, the Hadza organise themselves into bands of typically 10–20 people. They are a very ancient people, genetically distinct from all other ethnic groups. They separated from their closest relatives, the Sandawe, 15,000 years ago. Their language is completely unrelated to any other language. The Hadza have occupied their current territory for thousands of years, with little alteration in the way they live over that time. They have a rich oral history, with their own past being divided into four epochs.

THE HADZA PEOPLE OF TANZANIA, MUCH BELOVED BY ANTHROPOLOGISTS WHO HAVE STUDIED THEM IN DETAIL. THEY ARE A 'TRADITIONAL' PEOPLE, WHOSE LIFESTYLE AND CULTURE MIGHT REFLECT WHAT LIFE WAS LIKE FOR EARLY *HOMO SAPIENS*.

Their own 'origin story' is that at the beginning of time, the world was occupied by hairy giants, called Gelanebe, which means ancestors. The Gelanebe didn't have tools, couldn't make fire, hunted game by staring at it until it fell dead, ate meat raw, and slept under trees. In the second epoch, the Gelanebe were replaced by other giants, called Tlaatlanebe, who were less hairy. They could make fire, used medicines and could cast spells. The third epoch was inhabited by the Hamalwabe, which means 'nowadays'. They had weapons, built houses and had a gambling game called *lukuchuko*. The fourth epoch continues to today and is inhabited by the modern Hadza, who are called Hamaishonebe.

What about their encounters with others? There are many stories in the fourth epoch of Hadza women being captured by outsiders. Some Hadza women will marry into neighbouring groups, such as the Bantu Isanzu people, but they often come back to the Hadza tribe with their children. This may be a way of avoiding in-breeding. The Isanzu have been hostile to the Hadza, capturing Hadza men for many years, and even selling them into slavery in the 1800s. There have, though, been peaceful times between the Isanzu and Hadza, with intermarriage and periods of living together. In spite of this, the Hadza, since 1912, have been 'ready for war' with the Isanzu. Another nearby tribe, the Sukuma, have been on good terms with the Hadza, who allow them to drive their cattle through Hadza land in exchange for metal tools.

The interaction between the Hadza and neighbouring peoples is likely to be typical of how tribes have interacted with one another in Africa from antiquity and then on into the dispersal of humans out of Africa and throughout the world. If you encounter a stranger, you will be suspicious of them. Does this make racism and xenophobia a knee-jerk response? The idea of different races, which became common in the nineteenth century, first came about as a way to classify different humans, but also as a way to discriminate against others, particularly if they were seen as a threat. Tribes have always identified themselves as different from neighbouring groups. The idea of race as we define it today, meaning a grouping of humans based on shared physical or social qualities, came about as a consequence of Europeans coming into contact with groups from different continents. Skin colour and physical differences were used to group different people. In 1735 the taxonomist Carl Linnaeus divided the human species into four races: *europaeus, asiaticus, americanus* and *afer*.[4] Even then, one of the reasons to do this was to imply superiority. Linnaeus described *Homo sapiens europaeus* as being active and adventurous, while *Homo sapiens afer* was said to be crafty, lazy and careless. And so we see the emergence of racism as we know it today. Racism is defined as prejudice, discrimination or antagonism against someone of a different race based on the belief that our own race is superior.

In 1775 Johann Friedrich Blumenbach listed five races: the Caucasoid race, the Mongoloid race, the Negroid race, the American Indian race and the Malayan race.[5] Skin colour became a key defining feature, with the Caucasoid race (which was thought to have originated in the Caucasus – a region between the Black Sea and the Caspian sea, and a term that persists in the word Caucasian) being white, the Mongoloid race being yellow, the Negroid race being black and the American Indian race being red. From the seventeenth to the nineteenth century, these races were considered primordial, enduring and highly distinct. Many classifications were published, which concluded that other races were inferior to Europeans, particularly the Negroid race, and this was used to justify slavery. Perhaps surprisingly, even US President Thomas Jefferson considered Africans to be inferior to Europeans in terms of intellect, although he described Native Americans as equal to whites.

In the nineteenth century, the definition of different races was all about the subjugation of groups determined to be racially inferior. Many renowned scientists were of the view that different races had actually evolved separately in each continent and shared no common ancestor and had different hardwired traits. In the early twentieth century, anthropologists were of the view that race was a wholly biological concept, and that linguistic, cultural and social groupings existed along racial lines.[6] This became known as scientific racism and was used by the Nazis to justify a eugenics programme. The goal of the

JESSE OWENS (1913–1980) REALLY ANNOYED
HITLER BY WINNING FOUR GOLD MEDALS IN
THE 1936 OLYMPICS IN BERLIN, RIGHT UNDER
HITLER'S NOSE.

ESSE OWENS

EXTRAORDINARIO SPRINTER
GRO ESTADOUNIDENSE, CUATRO
CES CAMPEON OLIMPICO: EN
O Y 200 METROS LLANOS,

programme was to improve the so-called Aryan race, or as it was termed 'Übermenschen' or master race, which the Nazis believed they belonged to. The term 'racial hygiene' was also used. The Nazis believed that all other races were inferior. When US athlete Jesse Owens won four gold medals at the 1936 Olympics in Berlin, Albert Speer wrote that Hitler 'was highly annoyed by the series of triumphs by the marvellous coloured American runner, Jesse Owens. People whose antecedents came from the jungle were primitive, Hitler said with a shrug; their physiques were stronger than those of civilized whites and hence should be excluded from future games.'[7]

It has been found that race has no genetic basis, and so the term 'race' has been replaced with 'ethnic group' in the sciences. Nevertheless, the term 'race' remains as a sociological construct.[8] People have superficial differences (such as skin colour or a fold of skin around the eyes) but beneath the skin we are all highly similar, with differences within one ethnic group actually being more marked than between groups. Certain genetic variants may be evident in one ethnic group (e.g. alcohol intolerance is more common in Asians because of a deficiency in the gene for aldehyde dehydrogenase, an enzyme that breaks down alcohol) but this doesn't make them a different race. Racism actually means linking superficial differences, such as skin colour, with more extensive differences, such as being lazy. No such link exists.

All humans are classified as belonging to the species *Homo sapiens*. By the 1970s, by examining genetics, it was concluded that racial differences are mainly cultural, and anything that was not cultural (e.g. physical differences) was found in diverse groups of people at different frequencies. Human genetic variation occurs both within ethnic groups as much as between them. The genetic difference between humans is actually tiny, being of the order of 1–3 per cent. Following the human genome project, which mapped all of the genes in humans, the variation that was observed did not support the idea of genetically distinct races. Any differences have been shown to follow what geneticists term a 'cline' – a gradient that is observed in a trait across a geographical range.

If we take skin colour as an example, there is a cline from Northern Europe southwards around the Eastern end of the Mediterranean and up the Nile into Africa.[9] At one end, skin colour is pale white, and at the other it is dark, with skin colour getting progressively darker as we move up the cline. The same applies to most other physical traits, with no obvious boundary where a trait suddenly changes.

We now know that when humans left Africa, they all had dark skin, and then about 8,000 years ago, light skin became evident because of a genetic mutation, most likely in genes that have the names SLC24A5, SLC45A2 and HERC2/OCA2 and began to spread throughout the population.[10] The gene for SLC45A2 may have come from East Asian farmers coming into Europe about 5,800 years ago. This all happened because of people moving into environments with low ultraviolet radiation from the sun. People with light skin make less of the skin pigment eumelanin and this allows them to make sufficient Vitamin D in their skin from the weaker sunlight. Vitamin D is important for bone strength and also in the immune system. Asians owe their light skin colour to different mutations.[11]

Anthropologist Frank Livingstone has concluded that since clines cross racial boundaries, 'there are no races, only clines'. What all of this means is the term 'race' is now no longer acceptable. It is now seen as a social construct, a meaning put on something (in this case, humans being members of different races) by a society. The Council of the European Union went so far as to declare: 'The European Union rejects theories which attempt to determine the existence of separate human races.' In a survey of 3,286 American anthropologists carried out in 2017, there was a strong consensus that biological races do not exist.[12] Social scientists have replaced the term 'race' with 'ethnicity'. This refers to self-identifying groups based on shared culture, ancestry and history.

Even though the term is not grounded in science, the last US census, taken in 2010, still asked Americans to choose their race from options that included 'White', 'Black', 'Chinese' and 'American Indian'. For victims of racism, news

IRISH IBERIAN   ANGLO–TEUTONIC   NEGRO

The Iberians are believed to have been originally an African race, who thousands of years ago spread themselves through Spain over Western Europe. Their remains are found in the barrows, or burying places, in sundry parts of these countries. The skulls are of low prognathous type. They came to Ireland, and mixed with the natives of the South and West, who themselves are supposed to have been of low type and descendants of savages of the Stone Age, who, in consequence of isolation from the rest of the world, had never been out competed in the healthy struggle of life, and thus made way, according to the laws of nature, for superior races.

that the term 'race' has no scientific basis provides little solace. A poll carried out by the TV network NBC in 2018 found that 64 per cent of Americans say racism remains a major problem in American society and politics,[13] 41 per cent think that race relations are getting worse and 30 per cent think that race is the biggest source of division in the US. Four in ten African Americans say they have been treated unfairly

AN ILLUSTRATION FROM H. STRICKLAND CONSTABLE'S *IRELAND FROM ONE OR TWO NEGLECTED POINTS OF VIEW*, WHICH ILLUSTRATES THE APPARENT SIMILARITY BETWEEN 'IRISH IBERIAN' AND 'NEGRO' FEATURES, BOTH IN CONTRAST TO 'ANGLO-TEUTONIC' FEATURES.

in a store or restaurant because of their race in the previous month, 76 per cent say they experienced workplace discrimination based on their race;[14] and 51 per cent of all Americans think their president, Donald Trump, is a racist.[15] The recent Black Lives Matter protests in the USA, triggered by the murder of George Floyd, illustrate how serious the situation is.

On the other side of the pond, hooliganism provides us with an interesting sociological case study.[16] If ever there was a case of two tribes confronting each other it's when two sets of football fans, wearing their different colours, square off. There are countless examples of violence between football fans, which in England reached a peak in the 1970s. Football hooliganism has been studied by sociologists, partly as a way to come up with strategies to limit it. It involves a range of behaviour including taunting, spitting and fighting with or without weapons. It has been called 'ritualised male violence'. Involvement in football violence gives young men legitimacy, identity and power. Religion, ethnicity and class can all be part of it. In Scotland, the rivalry between Celtic and Rangers was partly religious – Celtic being Catholic and Rangers being Protestant. Football hooliganism became so common in England that it became known internationally as the 'English disease'. Fans of clubs such as Arsenal, Chelsea, Leeds United, Millwall and West Ham United were especially problematic. Actual racism became a key feature, with black players being taunted and having bananas thrown at them. Worryingly, in spite of campaigns to limit racism in football, the incidence seems to be on the rise. Reports of racism rose by 47 per cent in 2018/19 compared to the previous season.[17] Twenty-five per cent of footballers in the UK are black yet only 4 out of 92 managers are. Although there has been progress with stopping the violence, with measures such as all-seater stadiums and better security control, much still needs to be done to stem racism in football.

Part of the reason for racism is xenophobia. This is defined as a fear or hatred of anything foreign or strange. The word comes from *xenos*, the Greek for strange or foreign and *phobos*, meaning fear. UNESCO have made a distinction between racism and xenophobia, where racism is usually based on physical differences and xenophobia is more to do with an aversion to behaviour or culture. One of the first examples of xenophobia was in Greece itself, whereby anyone not from Greece was categorised as a 'barbarian' – someone to be hated and not trusted. Ancient Rome also held itself superior to other peoples, with Rome defining Macedonians, Thracians, Illyrians, Syrians and Asiatic Greeks as being 'the most worthless peoples among mankind and born for slavery'.

Nothing much seems to have changed. We see plenty of disturbing examples of xenophobia in today's world. In a survey in Canada, only 32 per cent of

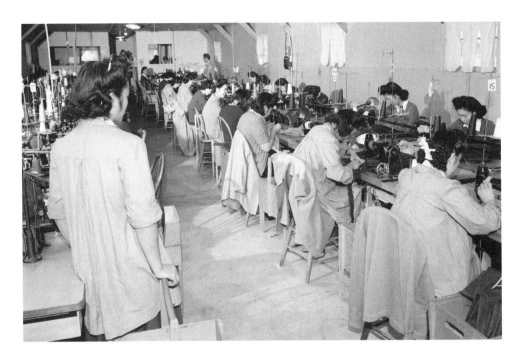

respondents said they had a 'generally favourable view of Islam'.[18] The US has a long history of xenophobia towards people from other countries: in World War II, in direct response to the Japanese attack on Pearl Harbor, they interned anyone who was Japanese[19] – 120,000 people of Japanese ancestry were interned in concentration camps; 62 per cent of those interned were US citizens. The internment is considered to have been due to xenophobia, rather than any security risk. Colonel

A PHOTOGRAPH TAKEN BY ANSEL ADAMS OF JAPANESE AMERICAN SUMIKO SHIGEMASU (LEFT) SUPERVISING FELLOW INTERNEES WORKING AT SEWING MACHINES IN THE MANZANAR RELOCATION CENTER, CALIFORNIA, WHERE JAPANESE AMERICANS WERE INTERNED DURING WORLD WAR II.

Karl Bendetsen, who was behind the programme, said anyone with 'one drop of Japanese blood' qualified for internment. People who were as little as 1/16th Japanese could be interned. The main administrator of the programme, General John DeWitt, repeatedly told newspapers 'A Jap's a Jap' and 'We must worry about the Japanese all the time until he is wiped off the map.' With remarkable philosophical resignation, the Japanese internees popularised the phrase 'shikata ga nai', which loosely translates as 'it cannot be helped'. Parents internalised their emotions to protect their children. The issue of Japanese internment became a running sore for Japanese Americans after World War II, and in 1980, President Jimmy Carter instigated an investigation into whether internment was justified. The report that emerged, entitled *Personal Justice Denied*, concluded that there was little evidence of any Japanese disloyalty and concluded that the justification was purely racist.[20] The Reagan government apologised and provided $20,000 in compensation for each survivor, with more than $1.6 billion eventually being paid out. The price of xenophobia.

Perhaps because the US is the great melting pot of different ethnicities, xenophobia is almost embedded in the US psyche. A network of civil rights and human rights organisations concluded that 'Discrimination permeates all aspects of life in the United States, and extends to all communities of colour.' Philosopher Cornel West concluded that 'racism is an integral element within the very fabric of American culture and society'. Russians are the latest target of xenophobia, with former Director of National Intelligence James Clapper on record as saying that Russians 'are almost genetically driven' to act deviously. President Donald Trump tried to enact a travel ban on people from Iraq, Iran, Somalia, Sudan, Yemen, Syria and Libya. Iraq was removed from the list when Trump was told that Iraq was an important ally in the fight against Islamic terrorism and that Iraqi interpreters, embedded within US forces, were being prevented from entering the US.

Xenophobia is also rife in other countries: 97 per cent of Lebanese, 95 per cent of Egyptians and 96 per cent of Jordanians are distrustful of Jewish people.[21] In 2012 the Human Rights Watch organisation concluded that Israeli Jews were racist against Muslim Arabs in 'institutional policies, personal attitudes, the media, education, immigration rights and housing'.[22] Meanwhile Ashkenazi Israeli Jews have been reported as holding discriminatory attitudes towards a whole range of other Jews, including Ethiopian Jews, Indian Jews, Mizrahi Jews and Sephardi Jews. Israel has implemented extensive anti-discrimination laws to try to combat racism.[23]

Harvard academics carried out a study in Europe from 2002 to 2015 that examined racism using data from 288,076 white Europeans.[24] The study used the Implicit Association Test – this is an accurate way to uncover unconscious bias or racism. It is a computer-based measure, which requires the subjects using it to rapidly categorise two concepts with an attribute. For example, in its simplest form, the subject will see 'Black' and 'White' and the word 'Unpleasant' and will be asked to choose which of 'Black' and 'White' would you associate with 'Pleasant'. Using this test for racial bias, the Eastern European Czech Republic, Lithuania, Belarus, Ukraine, Moldova, Bulgaria and Slovakia were most racist. Malta, Italy and Portugal also didn't do well in terms of racism.

South Africa gives us a vivid example of how a country handles the issue of racism and xenophobia. From 1948 until the early 1990s, South Africa had the most elaborate system of institutionalised racism ever seen. The system was called apartheid (from the Afrikaans *partheit*, meaning 'separate-ness'). Apartheid was a political system which ensured that South Africa was dominated politically, socially and economically by the country's minority white population. White citizens were legally granted a higher status, followed in descending order by Asians, Coloureds (meaning of mixed ethnic origin) and finally black Africans. It was split into two types: 'petty apartheid', which involved segregation of public facilities and social events and 'grand apartheid,' which was concerned with housing and employment opportunities. The first apartheid law was the Prohibition of Mixed Marriages Act of 1949, which was followed with the Immorality Amendment Act of 1950, which made it illegal for South Africans to marry or pursue sexual relationships across racial lines. In 1950 all South Africans were classified into the four racial groups,

NELSON MANDELA (1918–2013), ANTI-APARTHEID REVOLUTIONARY WHO SERVED 27 YEARS IN PRISON. HE WAS PRESIDENT OF SOUTH AFRICA FROM 1994 TO 1999.

A PROTEST FOR TRAVELLERS'
RIGHTS IN DUBLIN. IRISH
TRAVELLERS ARE AN ETHNIC AND
CULTURAL MINORITY IN IRELAND.

based on appearance, known ancestry, socioeco-
nomic status and cultural lifestyle. Where you could
live was determined by which group you were in.

Between 1960 and 1983, 3.5 million non-white South
Africans were moved into segregated neighbourhoods. Twenty-one thousand
people are estimated to have died in political violence during apartheid.
Generations of non-white South Africans have suffered from poverty and lack
of opportunity. Internal resistance led by such renowned figures as Nelson
Mandela (who was jailed for 27 years but subsequently became president of
South Africa from 1994 to 1999), Steve Biko (who was arrested in 1977 and
beaten to death by security forces while in custody) and international pressure
from everyone from the UN to the Catholic Church and most western countries
eventually led to the repeal of apartheid legislation on 17 June 1991. In spite of
apartheid being abolished, South Africa remains an unequal country. In 2018,
the World Bank called South Africa the most unequal country in the world
with accusations of corruption in the ruling African National Congress party
and growing economic and social inequality.[25]

Are Irish people xenophobic and/or racist? The Harvard study mentioned earlier indicates that the Irish are less racist than other European countries with an overall level similar to the UK. Yet there are clear concerns. Michael O'Flaherty, director of the Vienna-based EU Fundamental Rights Agency, has recently said that a four-year review of attitudes has found 'worrying patterns' of behaviour.[26] One in three migrants who were interviewed in 2019 said they had been discriminated against because of their skin colour – this is notably higher than the European average, which stands at 1 in 5. Seventeen per cent said they had suffered discrimination at work while 38 per cent said they had been subjected to harassment with 8 per cent suffering racist violence. Yet two-thirds of immigrants feel able to complain if they have been badly treated while 71 per cent say they had been treated respectfully by the police, which is better than in other European countries where the average was 59 per cent. Flaherty concluded that the sooner immigrants can be moved out of sheltered accommodation and into housing in the community, the better the integration will be. In 2005 Ireland was one of the first countries in the world to develop a National Action Plan against Racism, which ran for four years. The Irish government has instigated a number of initiatives, such as programmes to promote intercultural awareness and to actively combat racism and xenophobia. The success of these initiatives is up for debate.

These initiatives also extend to discrimination against the Traveller community. Irish Travellers are an ethnic and cultural minority, whose roots in the country go back centuries. Recent reports indicate that members are still being subjected to substantial discrimination in Ireland, despite government efforts to prevent racism and discrimination against them.[27] Genetic analysis has revealed that Travellers are ethnically distinct from other peoples in Ireland, with that distinction going back at least a thousand years, although a more recent analysis places the divergence from the settled community at somewhere between 240 to 360 years ago. This supports the idea that the Traveller community is descended from a group that began their nomadic existence following the final destruction of Gaelic society in the 1650s.

Travellers refer to themselves as Minkiers or Pavees and speak a distinct language called Shelta, also known as Cant, which is understood to be a mixed language derived from both Irish and English. Irish Travellers in the media have stated that they are living in Ireland under an apartheid regime. Irish journalist Jennifer O'Connell said: 'Our casual racism against Travellers is one of Ireland's last great shames.'[28] She acknowledges that there are problems with feuding and antisocial behaviour with some Travellers but to hold the entire community responsible is a typically racist response.

There are serious health issues in the Traveller community.[29] Only 3 per cent live to be 65. Suicide rates are six times higher than the settled community. Government attempts to help are failing to make a major difference. The issue of Travellers is seen as Ireland's last taboo subject (previous ones being homosexuality, contraception, divorce and abortion). In the last presidential election, a candidate called Peter Casey, who had made racist remarks about Travellers, came in second to Michael D. Higgins, disturbing many, although he performed badly at the 2020 general election.

What is especially disappointing is that discrimination against Travellers mirrors widespread discrimination against Irish emigrants in the UK and US. 'No Irish need apply' was commonly stated in both countries in the nineteenth and twentieth centuries. Prime Minister Benjamin Disraeli famously said: 'The Irish hate our order, our civilisation, our enterprising industry, our pure religion. This wild, reckless, indolent, uncertain and superstitious race have no sympathy with the English character. Their ideal of human felicity is an alteration of clannish broils and coarse idolatry.' More recently English journalist Julie Burchill said, 'I hate the Irish, I think they're appalling.' She narrowly escaped prosecution for incitement to racial hatred, following a column in the *Guardian* in which she described Ireland as being synonymous with 'child molestation, Nazi-sympathising and the oppression of women'.

How do we stop racism? Open, frank and objective discussions are necessary. Give and take is needed to counter the perceived threat of others, which can be easily stoked, with racist leaders preying on our suspicious and fearful

'NO IRISH NEED APPLY' BALLAD, WRITTEN IN 1878 BY JOHN
F. POOLE. 'SURE PADDY'S HEART IS IN HIS HAND, AS ALL THE
WORLD DOES KNOW / HIS PRATIES AND HIS WHISKEY HE
WILL SHARE WITH FRIEND OR FOE … THE DIVIL TAKE THE
KNAVES THAT WRITE "NO IRISH NEED APPLY".'

instincts. Any promotion of racial stereotypes will mean that people are set up to be racist on their first encounter with someone of a different ethnicity.

One strategy is to give the facts on how beneficial immigrants are to the local and national economy. When Trump tried to institute a ban on Muslims and Syrians coming to the US, he was reminded that Steve Jobs had a Syrian immigrant father. Former Taoiseach Leo Varadkar's father was an immigrant from Mumbai, India. Since 2000, 33 of 85 American Nobel Prize winners are immigrants. In the US, although immigrants make up only 13.7 per cent of the population in 2017, they made up almost 30 per cent of entrepreneurs. Forty-four per cent of the companies on the 2018 Fortune 500 list were founded by immigrants or the children of immigrants.[30] These companies brought in $5.5 trillion in revenue – an amazing return, bigger than the combined GDP of all the countries in the world bar the US and China.

Immigrants also own more than one in five of small businesses (which combined employ more people than big businesses). Immigrants contribute to the success of main streets in American towns and cities: Immigrants own 58 per cent of dry cleaners, 45 per cent of nail salons and 38 per cent of all restaurants.[31] They have helped stave off population decline in four out of five rural places in the US. They bring their food, music and culture. Without immigrants the working population of the US would be 7 million fewer. Even illegal immigrants play a big role in the US economy. In a recent study it was calculated that without them, the US annual GDP would be 2.6 per cent less.[32] Overall, immigrants are key to economic growth. They replenish and grow the workforce and are often highly qualified and skilled as well as being essential for establishing new businesses.

In spite of all this, in many countries discrimination and racism remain a major contributory factor for depression, ill health, low employment and wages, convictions and jail time. Black people are much more likely to end up in jail in the US than white people.[33] People of colour make up 37 per cent of the US population and yet make up 67 per cent of the prison population. Racism remains a major issue, both in the US and more widely. But there is hope: there is a global trend towards less discrimination on the grounds of skin colour. In the 1960s almost half of white respondents in the US said they would leave if a black family moved next door. This has fallen to 6 per cent.[34] In 1958 only 4 per cent of Americans said they approved of inter-racial marriage. Support is now at 87 per cent. Globally, young people are much less racist than their parents with 14 per cent of people under 30 expressing racist views, which compared to over 31 per cent in those over 30. Racism has to be seen as an instinctive response. The only way to deal with it is to encourage people to rise above it, for the sake of their fellow human beings. Perhaps most importantly we need to pay attention to those who have been marginalised. As Reni Eddo-Lodge has written, 'We must see who benefits from their race, who is affected by negative stereotyping of theirs, and on whom power and privilege is bestowed.'[35]

The bottom line: race is a social construct with no basis in science. We all need to keep learning about and correcting our prejudices, harnessing science as a weapon against discrimination while recognising that it has been used to justify horrors such as eugenics. There are differences between all of us, and these should be cherished. Racism remains an evil influence and we must all do our part to combat it.

CHAPTER II

WHY ARE YOU WORKING IN A BULLSHIT JOB?

—

'Lisa, if you don't like your job you don't strike.
You just go in every day and do it really
half-assed.'

—

Homer Simpson

AM LUCKY. I work in a job that I love. There are good days and bad days of course, but on balance the good outweighs the bad. How can you beat a job that combines teaching the next generation with being a scientist?

A recent survey in Ireland has revealed that 83 per cent of Irish workers think about quitting their job every day.[1] An astonishing 94 per cent of workers in China and Japan say they are not engaged in their work,[2] while 51 per cent of Americans say they have no meaningful connection with their jobs and tend to do a minimum.[3] Yet, another survey has revealed that 85 per cent of Americans say that they are either somewhat or very satisfied with their jobs (perhaps because they get away with doing little).[4] And what about that much-maligned group, the millennials? Well, 71 per cent are either not engaged in or are actively disengaged from their jobs.[5] All that education, hope, motivation, treks around the world to find themselves and youthful exuberance, and yet the majority are not happy in their jobs. Why is this? Why hasn't all the education, career guidance, expensive tuition, life coaching and mindfulness (for those who can afford it) not worked for most people when it comes to what they spend one-third of their time doing? Do people just complain anyway when they're asked, even when they might be somewhat happy in their work?

(Perhaps *The Matrix* got it right. In that movie, the character Agent Smith describes how the first Matrix was designed to be perfect, with no human suffering. It failed, and Smith concluded that humans define their reality through suffering and misery and so the programme is redesigned to accommodate that.)

For centuries, of course, there has been drudgery. The grunts who built Stonehenge or Newgrange or the pyramids were hardly happy, although it's possible that working in a group brought some satisfaction, given humans' sociable nature. The invention of agriculture seems to be partly to blame for the state of affairs, whereby there was a tiny minority lording it over the masses, making them work hard. There is now a consensus that agriculture gave rise to huge inequality.[6] The haves had the seeds, land, know-how and power to control agriculture and made the less smart or able or less well-connected people in their community work for them, and didn't share out the spoils equally. Slavery was invented to make sure the grunts kept working. The Romans knew that if the workers were given 'bread and circuses' (read gladiatorial combat), they would keep working and not rebel. Perhaps the equivalent of all this today is working in Facebook, being allowed to sit in nicely coloured chairs, eating a meat-free burger and watching a series on Netflix that involves the ritual humiliation of fellow human beings.

Things began to change somewhat when the idea of the Protestant work ethic began to emerge. Protestants, beginning with the man himself, Martin Luther, viewed work as a duty that benefited both the individual and society as a whole. The Catholics had spoken of 'good works', which meant helping your fellow human beings through your deeds and actions. These good works, combined with faith, ensured a one-way ticket to heaven. For Protestants, though, hard work was an indicator that you were 'elect', which meant you were predestined to go to heaven. Interesting idea to make people work hard, right? What this might have meant in practice was that even if you felt lazy, or tempted to do 'bad works' like, say, commit a crime – you could defy that because you were predestined to go to heaven. This might, in turn, have meant that you were less likely to hate your job as you were doing it as a precursor to going to heaven when you died. This might suggest that

MARTIN LUTHER (1483–1546), WHO KICKED OFF THE REFORMATION, WAS OF THE VIEW THAT WORK IS A DUTY THAT BENEFITS BOTH THE INDIVIDUAL AND SOCIETY, INSPIRING WHAT BECAME KNOWN AS THE 'PROTESTANT WORK ETHIC'. THIS HAS BEEN USED TO ANNOY LAZY CATHOLICS FOR CENTURIES.

unhappiness at work is a consequence of people not having faith in God.

Work–life balance refers to the balance that we apparently need between work and non-work things like hobbies and spending time with friends and family. It's seen as critical for our physical and mental health. There's no doubt that work is hard. You set your alarm clock to make sure you get up on time, make yourself presentable, commute in a crowded train, spend the day working – which for the vast majority means spending time in front of a computer screen – punctuated by meetings with people you either have to be with (because they decide your bonus) or actively dislike. You often get tied up in needless bureaucracy. You spend the day adding value to your company, and then you traipse home again, happy to be away from work. The statistics for who is doing what jobs bear out this description.

There are 400,000 office workers in the square mile of the City of London alone.[7] These are the people T.S. Eliot described in 'The Wasteland': 'A crowd flowed over London Bridge, so many, I had not thought death had undone so many.' Mind you, nowadays those people have earbuds in and are listening to all kinds of things to stop them from feeling dead. They are also holding almond-milk flat whites, which is perhaps the new comfort blanket. Two hundred million people in 40 developed countries currently work at desks in offices.[8] That's a lot of Post-it notes (if they still use such a primitive thing). The corporate office is the engine room of global economic growth. People commute to the office, spend around eight hours there, and commute home. Some life. (COVID-19 has, of course, changed this dynamic, with fewer people commuting and the interminable meetings being held on Zoom instead of in person. Progress! Or maybe not.)

It is intriguing to look at office design over the decades.[9] In the early twentieth century offices were all about efficiency – row upon row of typists and clerks. In the 1980s, big-haired office workers were stuffed into cubicles. Today, most offices are open plan, and many companies are moving towards 'hot desks', where no one has their own desk and instead sit at a different desk each day. You can't leave your favourite furry toy on it. Goldman Sachs recently moved 12,000 workers into a new $1.2 billion European headquarters in London.[10] It's all open plan, noise-reduction glass, hot-desking, dynamic air-conditioning and stairwells that optimise interactions between staff. They must now be wondering whether the design is safe in the wake of COVID-19. The British Council for Offices (yes, there is such a thing. I wonder where their offices are?) have estimated that average desk space has fallen by 10 per cent over the past nine years.[11] To compensate, employers provide ping-pong tables, climbing walls (where you can literally be driven up the walls) and pods you can sleep in should the need arise. The goal is to reduce workplace illness – on average, 10 per cent of a company's wage bill goes on sick pay.[12] Unilever recently calculated that every $1 spent on staff well-being results in an increase in productivity worth $2.50.[13] However, a study of 600,000 office workers revealed that 40 per cent felt that their office actually prevented them from working productively.[14] Given how much hot-desking is in vogue, the question is, does it work? It turns out that people don't like it. One study of a hot-desking office has shown that people spend on average 18 minutes trying to find a vacant desk, which adds up to 66 wasted hours per year.[15] They don't like having to clear their desks every night either. People are happier in surroundings that are familiar to them, marked by their own pot plant and novelty mug.

Whatever way you look at people's working lives, the bottom line remains the same: the majority find work hard and disengaging, and so we need to examine how to change that. It turns out it doesn't have to be this way, or at least you should aim to work at something you love, where the whole idea of life–work balance is nonsense. Did Neil Armstrong worry about work–life balance on his way to the moon? Mind you, he did say, 'The one thing I regret

A REPRODUCTION OF *GUERNICA*, WHICH WAS PAINTED BY PABLO PICASSO IN 1937 IN RESPONSE TO THE BOMBING OF THE BASQUE TOWN OF GUERNICA BY NAZI GERMANY AND FASCIST ITALY AT THE REQUEST OF DICTATOR FRANCO.

was that my work required a lot of travel.' Did Marie Curie worry about balance in her life when she made pioneering discoveries about radioactivity? And what about Picasso – did he feel like a long break after painting *Guernica*? I suspect not.

Not that everyone can be like these people, but we can at least strive not to feel that our working life is so awful that we feel tremendous relief when we are away from it. As comedian Bill Hicks said, 'There are two drugs that western civilisation tolerates: caffeine from Monday to Friday to energise you enough to make you a productive member of society, and alcohol from Friday to Monday to keep you too stupid to figure out the prison that you are living in.' Survey after survey confirms the crisis in the workplace. In another major study, carried out in 142 countries, the average level of engagement in work stood at 13 per cent.[16] A major reason turns out to be the latest *bête noire* of modern life: digital technology, which is exposing people to an unending flood of information they feel they have to respond to.

This leads us to consider what motivates us to work. Clearly, we need to make money to pay for things like food and shelter – our basic needs. But that is no longer a key reason for work. Many studies on happiness have shown that once we reach a certain level of income, we reach a level of happiness that won't be surpassed by earning additional income.[17] Some of us may have the drive to make a lot of money in order to buy a bigger house or car and lord it over our peers, but it turns out that money is not a significant motivating factor for many. We know deep down that such extra money will not necessarily make us any happier. And the idea of a universal basic income is gaining traction. This means the state will pay all its citizens a certain amount of money, which they can then spend in the economy. The idea actually goes back a long way: it was first proposed by Thomas Paine, the eighteenth-century radical who in 1797 suggested that all 21-year-olds be paid a £15 grant.

One motivation for a universal basic income nowadays is increasing automation. Mark Zuckerberg has argued that the increase in automation will create a greater need for a basic income for all. This especially applies to people currently working in low-paid jobs. A report to the US Congress in 2010 concluded that people in jobs earning less than $20 per hour were at an 83 per cent probability of losing their job to automation.[18] Some argue that if everyone were given an income, it would lead to increased consumption of alcohol or other drugs. In 2014 the World Bank commissioned a review of 30 separate studies into this issue, which concluded that this was not the case.[19] A post-capitalist society (which we may well be moving towards) would allow for a distribution of profits from companies that are publicly owned to the entire population, representing a return to each citizen on the capital owned by society. The Initiative on Global Markets Economic Experts panel at the University of Chicago has proposed that if every American citizen over the age of 21 were given a basic income of $13,000 per year this would bring all kinds of benefits to a society currently ravaged by inequality.[20] People with a basic income still get jobs in the economy and generate economic activity and growth. Advocates of a basic income argue that it will liberate workers from the tyranny of wage slavery and allow people to pursue different occupations, tap into their creativity, overcome the alienation they feel in their working lives and increase their leisure time.

But would it work? Finland is the most recent country to try it out. The Finnish government over a two-year period provided 2,000 unemployed people between the ages of 25 and 58 with a €560 per month salary.[21] It replaced unemployment benefit and people received it whether they got a job or not. One idea was that unemployed people might take on part-time work without fear of losing benefits. The Finns have long been seen as being at the cutting edge of social innovation. Their renewed interest in it stemmed from concerns about the increasingly fragmented nature of the labour market. Many people work in low-paid, low skilled work with growing wage inequality and unscrupulous employers getting away with exploitative practices. Social welfare systems are also becoming increasingly punitive. Predictions about the end of work are of course nothing new. In 1891 Oscar Wilde described a world where machines do all the work in his essay 'The Soul of Man Under Socialism'. Economist John Maynard Keynes predicted in the 1930s that people would have an average working week of 15 hours. So did the Finnish experiment work? Well, up to a point. Importantly it made people happier and healthier (surely an excellent outcome?), but it didn't make them more likely to become employed. This may have been because they were unemployed at the start of the experiment due to low skill levels or health issues. Yet many on the trial were positive about it, with one participant even writing two books.

A final interesting aspect of the idea of a universal basic income concerns the COVID-19 pandemic. Because of the mass unemployment that the pandemic brought, governments actually provided a basic income to many in order to stave off personal economic disaster, as well as social unrest. Who would have thought that it would be a virus, and not something like automation, that would have caused the first widespread example of a universal basic income in society? It's not clear yet what this has done for society, or indeed the economy, but it will be fascinating to find out.

If money is not a major motivating factor for us to work, then what is? This

brings us to Maslow's hierarchy of needs. In 1943 psychologist Abraham Maslow wrote a highly influential treatise
called 'A Theory of Human Motivation'[22] and followed it up

A PROTEST IN FAVOUR
OF A UNIVERSAL BASIC
INCOME FOR ALL.

in a book called *Motivation and Personality*. He used a pyramid to describe
what motivates us, with the biggest, most fundamental needs at the bottom
and the truly human, highest order needs at the top. At the bottom he put
physiological needs, which include food, water, warmth, sex and rest. We
are driven to seek these and without them we die. Next comes safety needs,
which means keeping ourselves out of harm's way and being aware of threats.
These needs protect us from injury. Next are what are termed 'belonging-
ness and love needs'. We are a social species and we like to be with others.
We also have an urge to procreate and need to love and be loved. Being in
isolation has severely negative effects on our mood, causing depression and
anxiety, although these aren't necessarily life-threatening. The lockdown
that happened during the COVID-19 crisis led to a lot of people being
isolated. It's not fully clear yet what negative consequences have happened
as a result of the lockdown because of isolation, but signs are that there was
a big increase in mental health difficulties for many. This need for belonging
occurs in most other social animals. Then we get to the higher-order needs,
which are largely specific to *Homo sapiens*. These include 'esteem needs'.

We crave the respect of others and a sense of accomplishment in what we do. Finally, we get to the highest need, 'self-actualisation': what this means is that we need to live a life true to our own interests and abilities. It can be achieved through things like parenting or in finding a mate. If you are a musician, you achieve this by playing music. If you are a mathematician, you achieve it through doing maths. You live a life that is true to your own calling. You fully realise your potential as a human being.

Maslow also wrote about self-actualisation leading to transcendence. You move beyond yourself into another realm, which can be thought of in a spiritual sense but can also involve altruism – you go beyond yourself. If any of these needs are threatened, we might either die (with the more basic needs) or be unhappy in our lives. I suspect few people achieve transcendence in the workplace ... but what motivates us to work is to meet the requirements of this hierarchy of needs. Work will provide us with the wherewithal to buy food, put a roof over our heads. It makes us feel secure. Job security is, in fact, something we all seek, and if that is threatened, it leads to unease. Losing one's job is a major risk factor for depression and anxiety. Being in a job usually involves working with others and so we fulfil our social needs. We also are granted esteem and accomplishment from work. Finally, we might actually achieve self-actualisation and hey presto, all our needs are met. Phew!

The sad thing is, for most people only some of these needs are met, and so it's no wonder many feel unhappy in work. There is data to back this up. In a recent study of 12,000 white-collar workers, employees were vastly more satisfied when their emotional and spiritual needs were being met – 'spiritual' in this context is the feeling of a higher purpose from their work.[23] This might relate to the discussion above about how some religions value work ethic and good works. It ties work into a spiritual dimension. The more effectively leaders and organisations met these needs, the greater the workers performed. In 2012 a meta-analysis, using 263 research studies across 192 organisations in 49 industries and 34 countries,[24] revealed that where staff felt engaged, there was 22 per cent higher profitability than companies who didn't have

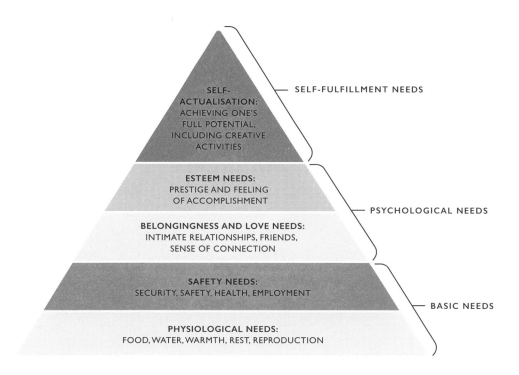

high engagement, along with 10 per cent higher customer ratings, 28 per cent less theft and 48 per cent fewer safety incidents. In simple terms, the way people feel at work has a huge effect on how they perform. A big negative was overwork. If employees take a break every 90 minutes, they reported a 30 per cent increase in their ability to focus. If people worked more than a 40-hour week they felt less engaged.

PSYCHOLOGIST ABRAHAM MASLOW PLACED THE THINGS THAT MOTIVATE US IN A PYRAMID. PHYSIOLOGICAL NEEDS WERE PLACED AT THE BOTTOM, AND HIGHER ORDER NEEDS ARE AT THE TOP, CULMINATING IN 'SELF-ACTUALISATION'.

A major reason for people not having their emotional and spiritual needs met at work is that they are working in what has been termed a bullshit job. This term was popularised in 2018 by anthropologist David Graeber in his book entitled *Bullshit Jobs: A Theory*.[25] Graeber contends that over half the jobs in our modern world are pointless. He describes five types of jobs that have no meaning but in which the workers pretend their job is not as meaningless

ANTHROPOLOGIST DAVID GRAEBER (LEFT) SPEAKING AT A CONFERENCE IN AMSTERDAM IN 2015. HIS BOOK *BULLSHIT JOBS: A THEORY* CLAIMS THAT HALF THE JOBS IN OUR MODERN WORLD ARE COMPLETELY POINTLESS. WHAT ABOUT PEOPLE WHO MAKE CHAIRS, DAVID?

as they know it to be: flunkies, goons, duct-tapers, box-tickers and taskmasters. Flunkies make their superiors feel important. Goons act with aggression on behalf of their employer (this category includes corporate lawyers and public relations people). Duct-tapers deal with preventable problems – an example here is the airline desk person who calms a customer down if their luggage doesn't arrive. Box-tickers use paperwork as a proxy for actual work. They include performance managers and in-house magazine writers. They spend a lot of time sending memos to each other in a never-ending cycle. Finally, the taskmasters are managers who create extra work for people with poor justification for the extra work.

Terrifyingly, as many as half of all jobs fit into these categories. If the people doing these jobs were lost from an organisation it wouldn't affect the productivity of the organisation. And worryingly, many people in these jobs know that the jobs are bullshit, either consciously or subconsciously. It reminds me of Golgafrincham Ark Fleet Ship B in Douglas Adams's *The Restaurant at the End of the Universe*. It was designed to relocate people from the planet Golgafrincham, which was being destroyed, to planet Earth, and was filled with what was termed the largely useless, redundant part of the population. In Adams's world he deemed these to be telephone sanitisers, account executives, hairdressers, tired TV producers, insurance salesmen, public-relations executives and management consultants. Ark Fleet ships A and C carried the rulers, or people who do useful work. Ship B was programmed to crash into the earth since it was full of 'useless idiots'. The other two ships are lost, but Ark Fleet Ship B makes it to Earth. The Captain can't remember why he has to crash the ship, but he does it anyway. After the crash, half the people die of a virulent disease contracted from a dirty telephone. The rest survive and continue their useless ways, giving rise to the Earth we see today. I wonder if this story is actually true?

Yet again, however, the COVID-19 pandemic has led to a change in our society in relation to how we view different jobs. Jobs that were perhaps thought to be not especially important suddenly became viewed as essential. This included jobs like hairdressers (up yours, Adams), delivery drivers, supermarket workers and cleaners. Jobs that were often overlooked or looked down upon. Perhaps Adams was being satirical in his list of people deemed useless, when the loss of telephone sanitisers led to disease from a contaminated phone that killed off half the people on Earth. Doesn't that sound familiar? His view of management consultants would probably not have changed much, however. If he were putting professions into Ark Fleet Ship B now, would he include hedge fund managers, insurance advisers and lifestyle gurus, who, in the post-COVID-19 world, are clearly useless.

Somewhat unexpectedly, most of Graeber's bullshit jobs are in the private sector, despite the widely held view that market forces should root them out. The constant cry of waste and inefficiency in the public sector is not necessarily justified. Graeber puts the phenomenon of bullshit jobs down to what he calls 'managerial feudalism' where underlings are needed to make managers feel important. Bullshit jobs also serve political goals: politicians only worry about numbers of people in jobs rather than what those jobs actually are. Following the publication of his book, a survey was carried out in the UK, which revealed that 37 per cent of people felt that their jobs had no meaning. Graeber is a keen advocate of a universal basic income to do away with bullshit jobs. He is of the view that increasing automation, instead of freeing us, has led to this state of affairs. Everyone knows it, including governments, but no one wants to do anything about it.

What can we do about this parlous state of affairs? One area that has been explored is remote working. Again, the COVID-19 pandemic brought this to the fore, with millions of people working from home during the lockdown. Before the pandemic, a study has revealed that in Europe, 70 per cent of professionals work remotely at least one day a week, while 53 per cent work remotely for at least half the week.[26] This is all a consequence of technology. Conference-call programmes such as Zoom are an increasing part of working life and became essential during the pandemic. All over the world right now, millions of people are dialling in, putting their phones on mute and waiting for their turn to speak. Is this the answer? No more commuting; freedom to organise your day as you see fit; getting the work/life balance right. People certainly seem to want it, with 70 per cent of millennials more likely to choose an employer who offered remote working.[27]

Sadly, though, surveys are showing that all may not be rosy in the remote working garden.[28] A 2017 United Nations report found that 41 per cent of remote workers reported high levels of stress, compared to 25 per cent of office workers. This is a concern to employers, where in the UK stress, depression and anxiety leads to £100m in losses every year. Why do remote workers feel stressed? One reason is 'out of sight, out of mind'. If you're not in the office you worry that you are being left out, that others are talking about you behind your back and that you are not trusted by your boss to

A DEPICTION OF WHAT IT MIGHT BE LIKE
TO WORK FROM HOME. THE PERSON WHO
DREW THIS PICTURE DIDN'T ANTICIPATE
THE REALITY, WHERE WORKING FROM HOME
MEANS SQUEEZING YOUR LAPTOP ONTO THE
KITCHEN TABLE, WITH LAUNDRY AND PESKY
CHILDREN A CONSTANT DISTRACTION.

be working hard. One study of 1,100 workers found that 52 per cent who worked from home even for some of the time felt left out, poorly treated, and were unable to deal with conflict with colleagues.[29] Navigating sensitive issues can really only be done face to face. Otherwise issues fester, emails are misinterpreted as rude and the absence of body language makes it difficult to communicate true meaning.

Another negative aspect of remote working is that when managers deal with remote workers, they are inclined to focus on tasks and not relationships. With an emphasis on deadlines, remote workers can feel like cogs in a big machine rather than an essential part of the team. Remote workers also sometimes have difficulty in switching off – their email is always on and they feel obliged to respond rapidly. (Interestingly, women were inclined to find remote working more stressful than men.) Crucially, physical face-to-face interaction is also a net de-stressor. A survey carried out by Google on what makes an effective meeting revealed something interesting – meetings in which the first 10–15 minutes were spent small-talking about the weekend, or how the kids are doing, were much more effective in achieving their goals.[30] Those colleagues who spend time in their busy day socialising (physically face-to-face, although it may well be that doing that online may work too) will be happier and therefore more productive.[31] This brings us back to Maslow: we have a basic need to socialise and remote working often doesn't meet this need. Remote working can work – but only if you check into the office physically from time to time.

Corporate leaders can have an influential role in defining the purpose of each job carried out in their organisation. When John F. Kennedy visited the NASA space centre in 1962, he noticed a janitor carrying a broom. He went over to the man and said to him 'Hi, I'm Jack Kennedy. What are you doing?' 'Well, Mr President,' the janitor replied, 'I'm helping put a man on the moon.' The key message here is no matter how large or small your role is, you are contributing to the larger story.

Then there is the fable of the three bricklayers, all working on the same

wall. Someone asks the bricklayers what they are doing. One says, 'I'm laying bricks,' another says, 'I'm building a wall' and the third says, 'I'm building a great cathedral for God.'

JOHN F. KENNEDY VISITING THE MERCURY SPACE CENTRE.

Which of these do you think felt most fulfilled in their job? The leader of the project is often the one who articulates the bigger purpose of the work being done and this can bring job satisfaction. If the leader can illustrate to the worker how their work is meaningful for the company then all the better.[32] A study of a call centre in the University of Michigan produced an interesting result.[33] The job was to cold-call alumni and ask for money. Yet when call-centre staff met a student who had received a scholarship in the programme, their weekly revenue increased by 400 per cent. In another example, when radiologists received a patient file that included that patient's photo, accuracy of diagnosis improved an incredible 46 per cent. And when workers who assemble surgical kits met the end users, they worked 64 per cent longer hours and made 15 per cent fewer errors.

It is also useful to match a person with a job they might find fulfilling. Career counselling has become a major growth area, with coaches (sometimes also called 'life coaches') advising on career choice and development. Career counselling has a long history, going back to the nineteenth century. Various approaches are taken to try and find where a person's true vocation might lie, from multiple-choice questionnaires, aptitude testing to essay writing to careful questioning and probing by the counsellor. Evidence suggests that it is useful, with counselling improving the chances of someone obtaining a fulfilling job by 32 per cent in one meta-analysis of 57 separate studies.[34] Counsellors can put qualifications and experience in a broader context, which might include desired salary, interests outside work and educational possibilities. So, if you can't decide how you might self-actualise in the workplace, or if you're in a rut in your job, go and see a career counsellor. At a minimum, you'll keep them in a job.

Lucy Kellaway is a great example of how to find the right job. Lucy is a journalist who worked for *The Financial Times* and used to write a regular column about the workplace. In 2016, after 31 years working as a journalist, she decided to leave and start a new career as a maths teacher. Having spent years slagging off the corporate workplace, she jumped ship. She co-founded a new charity called Now Teach, which aims to recruit experienced, mainly high-flying businesspeople, into teaching, particularly maths and the sciences. Now Teach is of the view that there is a crisis in teaching, and that helping people change profession to become a teacher will bring great benefits, both to the person themselves and also for the pupils, who will have a teacher bringing a lifetime's experience to the profession. Lucy retrained and took a job in a challenging London secondary school. Although

she loved journalism, she didn't want to spend her whole life doing it. She took to the new task with gusto but after a year said she had found it 'hell'. She found working full-time as a teacher extremely demanding, but when she switched to part-time and changed from maths to business studies and economics, she found that she 'really loves it'. She jumped from one career into another

LUCY KELLAWAY, FORMER JOURNALIST WITH *THE FINANCIAL TIMES* TURNED SECONDARY SCHOOL TEACHER, WHO WRITES A LOT ABOUT THE ABSURDITIES OF WHITE-COLLAR LIFE.

that she had always wanted to do, and then figured out how best to work in that new career. Lessons for us all in that.

Another lesson from Lucy[35] concerns the rewards she found in a particular job, as backed up by many books and studies on how to find meaning in the workplace. She came home from work one night and sat on a beloved antique chair, which had been left to her by an aunt. To her horror, the chair collapsed, and she fell on the floor. Because the chair was so important to her, she decided to try and fix it herself. It took her months. She had to obtain all the special materials. She tried to source materials from chairs of a similar vintage to maintain authenticity. She had to learn all kinds of new skills, from stuffing a chair with hessian to how to hammer in special nails to secure the leather seating. She had to learn how to varnish and preserve wood, and clear any woodworm from the old wood that was in the chair. It was tough and she almost gave up on occasion. At the end of it, she had fixed her chair, and had somewhere to sit. She felt a major sense of accomplishment and could enjoy her cup of tea.

So why did this activity, which involved such hard work, bring such satisfaction? She says there were three crucial elements. First, she had autonomy. She could decide when and where she would work on the chair. She could choose the material. Second, she had mastered a new set of skills. Mastery turns out to give us huge satisfaction. In the workplace, we will hopefully have gone through an elaborate education and mastered a set of skills whether by learning things of use or skills with our hands. Mastery is a key element of all jobs. And third, she had a purpose: to have something to sit on. This broader purpose brought a whole new dimension. It wasn't just about sanding wood or spending forever banging tiny nails into wood. It was about fixing a valuable thing. It was also about losing yourself in a task, which she found was often the case while fixing the chair. This is where we achieve what psychologists call 'flow'. It's an important goal for our overall mental health. When we achieve flow, we stop worrying about the kids, or the bills we have to pay, or why a colleague is trying to undermine us.

So there you have it. If your job has these three features: autonomy, mastery and purpose, you won't go far wrong. A key feature is for the worker to have some measure of control and power, otherwise they will be unhappy. (It must be emphasised, though, that even though people might feel unhappy in work, unemployment is far worse and engenders greater misery.)[36]

Luckily, my own job (although I don't call it a job) as an academic and scientist involves all three features. I can choose to spend my time among multiple different tasks, from teaching to scientific research. I have a degree in biochemistry and a PhD in pharmacology, which required mastery of both subjects. My purpose is clear: to inspire the young for their own and, hopefully, society's benefit, and to make meaningful scientific discoveries that might lead to new medicines for inflammatory diseases. This is why you'll sometimes find me on the night shift until 2 a.m., or spending endless time in airports on my way to meet other scientists or to give lectures. Or writing books like this – because I love to communicate science, too. **The bottom line: you can avoid or escape a bullshit job and lead a full and rewarding life. And won't that make your time on earth worth the hassle? And who knows, the COVID-19 pandemic might change our working lives and how we view different jobs forever – and for the better.**

WHAT YOU **LOVE**

PASSION MISSION

WHAT ARE YOU **GOOD AT**

IKIGAI

WHAT THE WORLD **NEEDS**

PROFESSION VOCATION

WHAT YOU CAN BE **PAID FOR**

IKIGAI IS A JAPANESE TERM MEANING 'A REASON FOR BEING'. IT'S A COMBINATION OF DOING SOMETHING THE WORLD NEEDS, THAT YOU LOVE, THAT YOU ARE GOOD AT, AND THAT YOU GET PAID FOR. IF AN INDIVIDUAL HAS IKIGAI, THEY WILL TAKE SPONTANEOUS AND WILLING ACTIONS WHICH WILL GIVE THEIR LIVES MEANING.

CHAPTER 12

WHY WON'T YOU GIVE ALL YOUR MONEY TO CHARITY?

—

'If you do good, people will accuse you of selfish
ulterior motives. Do good anyway.'

—

The Paradoxical Commandments by Kent Keith

GIVE TO CHARITY from time to time. Sometimes I'm guilted into it. Sometimes I do it because a friend involved in a particular charity asks me. Sometimes it even makes me feel good. But why don't I give away lots of my money, keeping just enough back for my family and me to live a modest life? I could do that, and yet I don't. Why not?

The fact is that the world is badly divided. Half of the world's net worth belongs to 1 per cent of the world's population.[1] Or how about this fact: the collective net worth of the world's poorest half (3.6 billion people) is equivalent to that of just eight of the world's wealthiest men – 8 = 3.6 billion.[2] Possibly the world's most unequal equation. Or this fact: the top 10 per cent of adults hold 85 per cent of all the wealth, with the other 90 per cent holding the remaining 15 per cent.[3] Nothing much seems to have changed over thousands of years, except perhaps during the occasional revolution. A tiny minority has held all the riches and lorded it over the peasants. How can this be? Why don't the top 1 per cent give up much of their wealth and spread the money around more evenly, especially when they know that many millions of people are suffering from poverty? And what about those of us who have more money than we need to maintain a good standard of living? Why don't we give up a lot of that wealth to others? Are we innately greedy or are we fearful of becoming poor or sick? Or is there something else going on? What can we do to redress this major inequality that exists in our world?

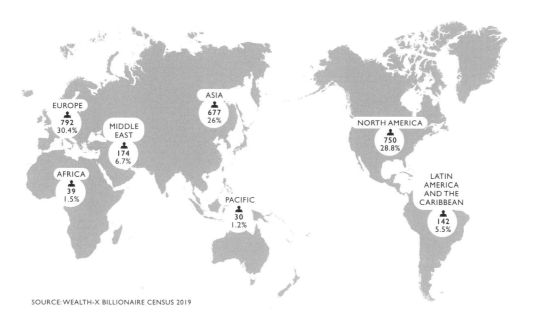

EUROPE
792
30.4%

ASIA
677
26%

NORTH AMERICA
750
28.8%

MIDDLE
EAST
174
6.7%

AFRICA
39
1.5%

LATIN
AMERICA
AND THE
CARIBBEAN
142
5.5%

PACIFIC
30
1.2%

SOURCE: WEALTH-X BILLIONAIRE CENSUS 2019

Let's examine the numbers a bit more closely. First, the super-wealthy. According to the Wealth-X Billion-aire Census 2019 (now there's a list to be on), there are currently 2,604 billionaires in the world.[4] A billionaire has a net worth of at least $1 billion – a thousand million dollars. Over 25 per cent of them live in the US, which hosts 705 of them. China has the next biggest proportion with 285. Germany comes next with 146. The report has found that since 2018, the combined wealth of billionaires dropped by 7 per cent to a measly $8.6 trillion. Some of these people are so rich that a new number had to be coined to capture how much money they have; Jeff Bezos, who founded Amazon, hit a net worth of $112 billion in 2017 and is now known as a 'centibillionaire'. And all because he invented a delivery service. He is currently the richest person in the world, overtaking Microsoft's Bill Gates, who held that title for a number of years. However, he's not the richest person of all time. That title goes to oil magnate John D. Rockefeller, who became a US dollar billionaire in 1916, which makes him history's wealthiest

NUMBERS OF BILLIONAIRES IN DIFFERENT PARTS OF THE WORLD, AS OF 2019.

JOHN D. ROCKEFELLER (1839–1937), AMERICAN BUSINESSMAN AND PHILANTHROPIST. HE IS WIDELY REGARDED AS THE WEALTHIEST AMERICAN OF ALL TIME. HE WAS THE FIRST MODERN PHILANTHROPIST, FUNDING MEDICINE, EDUCATION AND SCIENTIFIC RESEARCH.

person if we adjust his wealth to today's money.

Ninety per cent of billionaires are men, but the trend for billionaire women is on the up. In the last five years, the number of female billionaires has grown by 46 per cent, which is more than the growth in male billionaires in that period, which stands at 39 per cent.[5] There are now 233 female billionaires in the world, up from 160 in 2013. Only 11 billionaires are of black African ancestry, the wealthiest being Aliko Dangote, a Nigerian businessman with multiple business interests, who is worth $8.9 billion. Oprah Winfrey is also on the list with a net worth of $2.7 billion.

There are a lot of metrics when it comes to measuring billionaires, and it is a topic of endless fascination for many. I wonder why? One interesting aspect is which universities produce the most billionaires.[6] The US again dominates this list: Harvard has provided 188, followed by Stanford with 74. In fact, all of the top ten billionaire-producing universities are US-based. It's no coincidence that 39 of the top 100 universities in the world are in the US, according to The Times Higher Education World University Rankings 2020. Graduates in the US give a lot of money back to their universities. In 2018 Harvard received $1.42 billion in donations. Stanford came next with $1.10 billion. In 2018 Trinity College Dublin announced a donation of €25 million from the Naughton family, the single largest philanthropic donation in the history of Ireland. Irish universities need more Naughton families.

As of 2019 there are nine billionaires in Ireland.[7] Number one is Pallonji Mistry, an Indian-born Irish construction tycoon with a net worth of $14.4 billion. He became an Irish citizen after marrying Irish national Patsy Dubash, and he lives in Mumbai. Hilary Weston is at number two and is worth $8.6

billion. She was born and raised in Dublin and married Canadian Galen Weston in 1966, who made his money in the food industry. American-born John Grayken comes in at number three and is worth $5.9 billion. He is chairman of private equity firm Lone Star Funds and in 1999 became an Irish citizen for tax purposes. (Welcome, John, and make sure to pay your taxes.)

MARTIN AND CARMEL NAUGHTON HAVE MADE NUMEROUS PHILANTHROPIC DONATIONS TO MANY CAUSES, INCLUDING SUBSTANTIAL DONATIONS TO TRINITY COLLEGE DUBLIN. IN 2016, HE AND CARMEL WERE NAMED PHILANTHROPISTS OF THE YEAR BY THE COMMUNITY FOUNDATION OF IRELAND.

As for millionaires, there are just over 47 million of them in the world, with the US again dominating with 18.6 million millionaires.[8] This number is difficult to estimate, though, as it includes assets such as property. In 2019 the number of Irish millionaires rose to 78,000, an increase of 3,000 on the previous year mainly due to increasing asset and property values.[9] Of these, 1,029 have more than $30 million, which places them in the 'ultra-high-net-worth individual (UNHWI)' category.

So where did these people get their money?[10] Globally, for men, 62 per cent are self-made, which mainly includes entrepreneurs who set up their own businesses, 7.9 per cent inherited their wealth and 30.1 per cent are a combination of the two. For women, the difference is remarkable – 16.9 per cent are self-made, 53.3 per cent inherited their wealth, and 29.6 per cent are a combination. These percentages will change as we see more female entrepreneurs enter the list. As we move down the scale to millionaires, we see that there is a range of reasons for how people manage to accumulate wealth. In western countries, entrepreneurship has been shown to account for three-quarters of new millionaires. Other ways to become wealthy include pursuing a career with the end goal of becoming what is called a C-level executive in a company, such as a Chief Executive Officer (CEO), Chief Operating Officer (COO) or Chief Financial Officer (CFO). People who become a leading professional in a specific field or a top corporate salesperson often become millionaires. Only around 1 per cent of new millionaires become wealthy by other means, such as sports, show business or in the arts. It is rare to become wealthy in these fields of endeavour. So, it can be concluded that becoming a billionaire is extremely rare; becoming a millionaire less so, but it is still uncommon in the general population.

Now let's examine people giving their money away. We enter the world of philanthropy. This is defined as the desire to promote the welfare of others, especially by the generous donation of money to good causes. Another definition is private initiatives for the public good, focusing on quality of life. It is somewhat different from charity, which aims to relieve specific social concerns. Philanthropy seeks to address the root cause of a problem. The difference is captured in the well-known metaphor of giving a fish to a hungry person as opposed to teaching him how to fish. Philanthropists generally teach a man to fish, if you know what I mean.

First, the good news. There has been an upward trend in philanthropic activity over the past decade, which correlates with the increasing number of billionaires. The Wealth-X Billionaire Census has concluded that this is also due to increased awareness of global environmental and social issues, 'consternation over rising inequality' and a more diverse and multi-generational billionaire population. Nice to read that some billionaires, at least, are feeling consternation. The numbers show that globally the top 20 billionaires donated 0.8 per cent of their total wealth in 2018.[11] Not much, is it? Some give discreetly for personal, cultural or religious reasons. More than half, though, are involved in philanthropic giving either through organisations that they themselves have established or by other means. Thirty-five per cent of them have their own charitable foundations.

When we examine where the money goes, we see some interesting trends.[12] Education comes out tops. Two-thirds of billionaires give money to scholarships, educational support, outreach programmes and teacher training, and, in total, 29 per cent of all billionaire donations go to education. This is because it's a major way in which education is funded in the US. Healthcare comes in next, with 14 per cent of donations. Ten per cent goes to arts, culture and sports, and 8 per cent is donated to environmental issues (one has to wonder if that figure will increase with the recent major increase in awareness of climate change). Finally, 5 per cent goes to religious organisations.

An interesting development in the world of philanthropy occurred in 2010 when Bill and Melinda Gates and Warren Buffet (then the world's number one and number two in terms of wealth) started the Giving Pledge initiative.[13] The aim of this campaign was to get wealthy people to donate at least half their wealth, in their lifetime or in their will, to charitable causes. Initially, 40 people signed up, all in the US. As of April 2020, the number rose to 209, from 23 different countries. This translates into a total pledge of $600 billion by 2022 – not bad going, provided it happens.

According to its website, the Giving Pledge was inspired by the example set by 'millions of people at all income levels, who gave generously – and often at great personal sacrifice – to make the world better'. This is an interesting statement as it suggests that what Gates is trying to do is guilt the hugely wealth into giving more. If poorer people can do it, then you should too. The organisation aims to 'shift the social norms of philanthropy among the world's wealthiest people and inspire people to give more'. The

BILL AND MELINDA GATES AND WARREN BUFFETT STARTED THE GIVING PLEDGE INITIATIVE, WHICH AIMS TO CONVINCE THE WORLD'S WEALTHIEST PEOPLE TO GIVE HALF OF THEIR WEALTH IN THEIR LIFETIME TO CHARITABLE CAUSES. BY 2020, OVER 200 PEOPLE HAD JOINED THE PLEDGE, WHICH WILL RESULT IN A TOTAL DONATION OF $600 BILLION, PROVIDED THE BILLIONAIRES ALL COUGH UP.

pledge recognises that the challenges the world faces are highly complex and need input from governments, non-profits, academic institutions and businesses. It aims to act as a catalyst to promote investment in areas that governments and business can't or won't fund. But it has its critics: it's a 'moral' rather than a 'legal' pledge, so no one is actually obliged to give any money. Attention has also been focused on the size of the pledge rather than what it will go towards. It's yet another attempt to convince rich people to part with their money for the greater good, and it may well work.

Encouraging people to give money to charitable and philanthropic causes has, of course, a long history. The word 'charity' originated in late Old English and meant 'Christian love of one's fellow'. It, in turn, comes from the Latin '*caritas*', which describes a particular form of love for your fellow humans. Charitable giving was an act or duty in several religions and was referred to as almsgiving or tithing. A tithe was defined as one-tenth of something (usually income but also goods), to be given to a religious organisation or government. Traditional Jewish law included tithing, and Orthodox Jews still give one-tenth of their income to charity, as do Mormons. In Christianity, Jesus taught 'tithing must be done in conjunction with a deep concern for justice, mercy and faithfulness'. The fact that 10 per cent was a suitable amount to give is interesting – perhaps not too much to put people off or encourage them to evade giving, and not so little that it wouldn't make much difference to the cause being supported. In medieval Europe during the twelfth and thirteenth centuries, rich people built hospitals for the sick and poor. Religious orders were founded, whose main mission was to engage in charitable works. The first children's charity was established in England in 1739 by Thomas Coram of the Foundling Hospital, which looked after 'unwanted orphans'. Another noted philanthropist of that era, Jonas Hanway, established The Marine Society, for seafarers down on their luck, and the Magdalene Hospital, which aimed to help prostitutes. In the nineteenth century we see the emergence of campaigners, notably William Wilberforce, who championed the cause of the abolition of slavery. By 1869 there were over 200 charities in London, with a combined income of £2 million.[14] A sample of 466 wills in the 1890s revealed a total wealth of £76 million, of which £20 million was left to charities.

Philanthropy starts to take off with a vengeance from the late 1800s, a notable example being the Guinness Trust. It was founded in 1890 by Edward Guinness, 1st Earl of Iveagh, the great-grandson of Arthur Guinness himself. He donated £200,000 – the equivalent of £25 million in today's money.[15] This initiative included The Iveagh Trust in Dublin, which aimed to provide affordable housing in and around Dublin. As of 2018 the Guinness Trust owns and manages around 66,000 homes and provides services to more than 140,000 people in the UK and Ireland.[16]

THE IVEAGH TRUST WAS ESTABLISHED IN 1890 BY EDWARD GUINNESS, FIRST EARL OF IVEAGH, WITH A DONATION OF £200,000.

How much is currently being given by philanthropy? The US has the best statistics on donations.[17] In 2018 Americans gave $427.71 billion, an increase of 0.7 per cent on 2017; $290.84 billion of this came from individual philanthropists, followed by $75.86 billion from foundations, with $39.71 billion coming from bequests. Corporate giving in 2018 was $20.05 billion, a 5.4 per cent increase on 2017. There are currently 1.5 million charitable organisations in the US. In Ireland, registered charities (which number around 10,000, and include hospitals and universities) raised €14.5 billion.[18] Government and public bodies are the primary source of this income, accounting for €7.7 billion of funds raised. Hospitals received €3.1 billion and universities just under €3 billion. Because the government is the main source of funding, this doesn't really count as philanthropy. Half of the registered charities had an income of less than €250,000.

Historically, the largest philanthropic donations in Ireland by far have been made by Atlantic Philanthropies,[19] a private foundation created in 1982 by Irish-American businessman Chuck Feeney. Its main goal is to provide funding for health, as well as socially and politically liberal causes. In 1982 Feeney transferred all his assets and the entire ownership of his company, Duty Free Shoppers, to Atlantic Philanthropies. For the first 15 years of its existence, these donations remained anonymous. The total amount of money given by Atlantic Philanthropies to date is $7.5 billion – that is a remarkable amount to come from just one person. Over $1 billion was invested in third-level education in Ireland, including major donations to the University of Limerick, Dublin City University and Trinity College Dublin. This funding had the effect of leveraging a further $1.3 billion from the Irish government for the third-level sector in Ireland – a striking achievement. The investments are considered an important part of the stimulus that happened in the Irish economy in the 1990s. That one philanthropic foundation could have such an impact on a whole country is an example for others to follow. Other funding included a donation of $11.5 million for political advice to the Gay and Lesbian Equality Network. In 2015 Atlantic donated $177 million to University

of California San Francisco and Trinity College Dublin to create the Global Brain Health Institute, which aims to stem the rise in dementia by connecting researchers and physicians in this area.

CHUCK FEENEY, AN IRISH-AMERICAN BUSINESSMAN AND PHILANTHROPIST. HE FOUNDED ATLANTIC PHILANTHROPIES, WHICH HAS DISTRIBUTED MORE THAN $7.5BN.

The big question that arises from philanthropic donations (and indeed all charitable donations) is how to determine the impact of the donations being made. Atlantic Philanthropies may be one of the best examples, where a clear impact on the Irish education system can be seen, which has paid dividends for Ireland and indeed the rest of the world. A recent analysis has concluded that philanthropists are effectively 'flying blind'. An example of this is Facebook co-founder Mark Zuckerberg's $100 million donation towards building new schools in Newark, New Jersey, which has been accused of achieving little – $20 million of the $100 million went to consultants.[20] Teachers and parents became resentful over what was being proposed and the end result was *less* investment in public schools in Newark, with teachers actually being laid off. So what has been proposed is a 'science of philanthropy' to examine the best way for philanthropists to fund projects or initiatives.[21]

The Centre for Effective Philanthropy, which is based in Cambridge, Massachusetts, has recently reported that the time spent on preparing and then managing ten grants of $10,000 each is six times as long as the time spent on one grant of $100,000.[22] The London-based consultancy firm nfpSynergy examined UK charities and found that the charities value £2 of unconditional funds (meaning that the money is given with no strings attached) as the same as £3 of conditional funds (meaning that the money is given to a charity with conditions attached). The Shell Foundation has found that three times as many of its grants succeeded when it was involved in creating and managing the work, rather than when it came from someone who sent in an application for funding. Another important question for funders is how big a grant should be. A study on the impact of arthritis research found that large grants were no more effective than small ones. This may be because smaller grants fund a more diverse range of projects.

The unrelenting question if you're a philanthropist becomes: how do I know my money is achieving the goal I want? To give them credit the eight wealthiest men whose wealth equals the 3.5 billion poorest people are actually mega-philanthropists, funding billions of dollars' worth of medical research, public health, education and a range of humanitarian causes. They include Bill Gates, who as of 2019 had donated $35.8 billion worth of stock in Microsoft to the Gates Foundation,[23] Mark Zuckerberg, who with his wife Priscilla Chan has pledged to give 99 per cent of their Facebook shares (which at current market value equals $48 billion) to the 'cause of human advancement'[24] and Jeff Bezos, who gave an estimated $2 billion to launch the Bezos Day One Fund, which aims to help homeless families and support pre-schools in low-income communities.[25] Bezos has yet to join the Giving Pledge. Health issues are squarely in the sights of these mega-rich people. A large proportion of the Gates money has been spent on vaccine development and global health. Zuckerberg has set aside $3 billion to 'cure, prevent or manage' disease.

Michael Bloomberg, American businessman, politician, philanthropist, and author has allocated $1 billion to decrease smoking and traffic deaths.

One interesting development is the age at which the mega-rich become rich. It used to be old guys who gave their money away, be it John D. Rockefeller or Andrew Carnegie. But now we have Mark Zuckerberg, whose donation will exceed that of Rockefeller, Ford and Carnegie *combined*. Zuckerberg is 35 years old. John D. Rockefeller only set his foundation up when he was 76, and Andrew Carnegie opened his first library when he was 68. All of this seems good: lots of money being invested into good causes, and quickly. But there is criticism, especially in the US where previously, governments used to collect billions from the wealthiest people and then democratically redistribute the money. This is less and less the case. We now have a scenario where the wealthiest pay much less tax than before and then donate their money as they see fit, in an essentially undemocratic manner. There is no clear way to assess whether the money is being spent effectively or on the right things, which has led to a call in the US to rein in the mega-billionaires and go back somewhat to the old way of doing things: tax them and redistribute the wealth that way.[26]

And yet there are remarkable success stories, such as Atlantic Philanthropies and the Irish education system. Another is a project funded by the Gates Foundation in 2001,[27] which gave $70 million to the global vaccine organisation PATH and the World Health Organization over a ten-year period in order to develop a vaccine for meningitis A and make it affordable for everyone who needs it. A successful vaccine was developed at a cost of 50 cents per dose and by 2013, less than a decade after a meningitis outbreak had killed 25,000 people (mainly young people), four cases have been reported. The reason

JEFF BEZOS HAS MADE SO MUCH MONEY THAT A NEW TERM TO DESCRIBE HOW RICH HE IS HAD TO BE INVENTED: CENTI-BILLIONAIRE.

this succeeded is because of the sustained funding (there are many setbacks in research and long-term funding is key) and a hands-off approach by the Gates Foundation. Give money to the experts and let them get on with it.

Whatever about the high-net-worth individuals giving money away, what about the general public? A lot of work has gone into examining what it is that people need in order to give to charity and reasons why they don't.[28] Understandably, the first concern people have is to themselves and their own families. People might worry about saving money for their children's education (especially in the US) and, increasingly, their own retirement. In Ireland, a recent survey demonstrated that over 80 per cent of people worry about their own financial security in retirement.[29] Only one in six are confident about being financially sound when they retire. This is a higher level of confidence than in other European countries, which could be because of the relatively high state pension in Ireland. This currently stands at €238 per week, as opposed to €126 per week in the UK, for example. Fear of financial troubles in retirement impedes people from giving to charity.

People also think that any donation they make will be too small to make a difference, or that the money they give will be misused in some way. Charities make a great effort in stating how much bang donors will get for their bucks. In 2018 an analysis was made of those charities that benefit the most from donations.[30] Charity recommenders such as GiveWell, Charity Navigator and GuideStar rate charities in terms of which of them will use the money you give most effectively and who has the highest funding needs. Top of the list was the Malaria Consortium, which helps provide anti-malarial medication to children. Malaria is seen as a key issue to solve in Africa because it is the single biggest drag on the African economy. Why was that particular charity singled out? It has a clear goal: to distribute preventive anti-malarial drugs to children age 3–5 years during the high transmission season. There is substantial evidence that this charity makes a major difference, with 95 per cent of children receiving at least one month's treatment. The charity specifies that $6.80 provides four months of treatment to a single child. GiveWell were also of the view that further investment would have a massively beneficial effect on the Malaria Consortium. Second on the list was the Against Malaria Foundation, which provides insecticidal bed nets in Africa and Papua New Guinea: $4.59 will buy an anti-malarial bed net, which will protect two people for three years.

In 2016 *The Irish Times* analysed where money given to various Irish charities ended up.[31] Charities had been getting something of a bad press in Ireland. Misappropriation of funds by staff had become a concern and Charities Institute Ireland was launched in 2016 to promote best practice and restore public trust in the sector. A key issue was how much of the money people gave actually went towards direct activities. The charity ALONE, which helps older people, was found to give 100 per cent of what was donated to frontline services. Oxfam Ireland was found to put 80 cents of each euro donated towards easing poverty around the world. Concern, which also fights poverty, was also found to perform well, with 91.1 cent of every euro donated going directly towards charitable works. Other charities didn't perform as well using this metric. In the UK, a similar and rather damning report revealed that of more than 5,000 charities surveyed, one in five were spending less than 50 cent per euro on charitable work.[32] This metric has been criticised, though, as some charities have to spend a proportion of their income on fundraising efforts – this can include running shops, which comes with an overhead.

Global charities that aim to help relieve poverty in the developing world face many challenges.[33] People often feel it's the responsibility of governments more than charities. This led the United Nations to launch, in 2000, the United Nations Millennium Development Goals, which asked the governments of all developed nations to allocate 0.7 per cent of gross national income to overseas development assistance.[34] Only five countries have reached this goal: Denmark, Luxembourg, the Netherlands, Norway and Sweden, with the UK and Finland almost hitting the target. The argument goes that individuals in each country can make up the difference by donating. A number of goals were established and there has been progress on several, although this varies from country to country. A major goal reached was to halve the proportion of people in the world living on less than $1 per day, which was achieved in 2008 and continues to see improvement. Since 2015, the aims have been refocused and gender equality is now a key underpinning objective for all of the goals. Other barriers to people giving to global charities include a perception that aid makes countries depend on foreign resources and that giving leads to overpopulation. Arguments are made against both of these points, with a partnership model between charities and governments being important (for example, the Bill and Melinda Gates Foundation often leverages government funding when it invests in projects) and clear evidence that economic development actually lowers the birth rate. Educating girls to as late an age as possible has been proven to lower the birth rate.[35]

To turn the question around: why do people give? A major recent study has pulled together 500 studies in order to examine the key factors that drive giving.[36] For 85 per cent of donations, the main reason for giving was 'I was asked.' This reason might seem obvious enough, but it doesn't answer why the donor chose to say yes to a specific charity. Most people give to affirm important personal values, which include compassion to those in need. Donors also report that giving makes them feel good, or makes them look good in the eyes of others. In one study of 819 Americans who had given to charity in the previous month, five main reasons emerge, which have been given the acronym TASTE.[37] First, trust: the donor has to trust the charity they are contributing to. Second, altruism: the need to help others. Third is social: donations should matter to someone the donor knows and cares about – for example, you might know someone who has a disease, so you support research into that disease; or someone invites friends to a fundraising event and there is a collective donation. Tax is fourth: if people get a tax break, they are more likely to give. Many governments provide tax relief for donations to charities. In Ireland, the Charitable Donation Scheme allows charities to claim back tax paid on a donation of over €250. The fifth and final reason is egotism: people want to feel a personal benefit, say, by looking good in the eyes of others. Overall, though, people are more likely to be motivated by helping others rather than getting something back.

What might the future for philanthropy and charitable giving look like? The trend for philanthropy looks good, with an upward trend evident, but this will depend on the state of the global economy. Yet again, COVID-19 might have changed things. The billionaires of the world are giving lots of money to projects related to COVID-19. According to *Forbes* magazine, 77 of them gave often undisclosed but substantial amounts.[38] Jack Dorsey, CEO of Twitter, gave $1bn. Bill Gates committed $105 million. Even Donald Trump has given $100,000. They appear to be stepping up.

Yet for charity overall, some disturbing trends are emerging. In the UK in 2019, the proportion of people saying they had given to charity had dropped from 61 per cent to 57 per cent over a three-year period.[39] When this was investigated, one reason was a decrease in charities directly asking for donations because of GDPR legislation lowering the rates of direct mailing, which charities had heavily relied upon. On a brighter note, there was an increase in legacies and community fundraising initiatives. Digital engagement is also on the rise, which should help. Giving will fluctuate but will hopefully continue and grow in the future.

Through history there have been numerous critiques of charity, which persist today. One thing Oscar Wilde was well known for (other than his poetry and plays) was an essay he wrote entitled 'The Soul of Man Under Socialism'. In it, he calls charity 'a ridiculously inadequate mode of partial restitution, usually accompanied by some impertinent attempt on the part of the sentimentalist to tyrannise over the poor's private lives'. He was of the view that charity prolongs the 'disease' of poverty, rather than curing it. Oscar got it wrong, although perhaps one of his goals was to urge governments to do more. What else should rich people do with their money other than help humanity? And the same applies to all of us. **Bottom line: give what you can and pay your taxes as an act of philanthropy. As a famous good guy once said, 'For it is in giving that we receive.'**

CHAPTER 13

WHY ARE
YOU WRECKING
THE PLANET?

—

'If you really think that the environment is less
important than the economy, try holding your
breath while you count your money.'

—

Guy McPherson, Professor Emeritus of Natural Resources
and Evolutionary Biology, University of Arizona

LIKE MOST HOUSES, ours has three bins. A black one. A green one. And a brown one. It's complicated. So complicated that your professor here sometimes puts the wrong thing in one of the bins. Tension ensues. The bins are part of the battle to save the planet.

The story of how we are wrecking the planet began millions of years ago when microscopic organisms, creatures called zooplankton and algae, died in their billions as part of their natural life cycle and settled at the bottom of the sea.[1] They were then slowly covered up with silt and mud over millions of years. Under the weight, they began to decompose and form a waxy substance which is called kerogen. This is mainly made of chemicals called hydrocarbons. Things then got hotter down there and slowly, the kerogen turned into a liquid called petroleum. Millions of years later, if you were lucky enough to live over these lakes of underground petroleum, you would become enormously rich. Because petroleum became one of the most sought-after things on earth. How did that happen and why is the burning of the liquid remains of these ancient creatures killing the earth? And what can we all do about it?

The first written account of the use of petroleum was by Herodotus, an ancient Greek historian who lived c. 484–425 BC.[2] He described how a semi-solid form of petroleum called asphalt (also known as bitumen) was used in the construction of walls and towers in Babylon. There were vast quantities of asphalt on the banks of the Is river, a tributary of the Euphrates. The Chinese also wrote about petroleum and provide the first record of its use as a fuel, as early as the fourth century BC. The Japanese described it as 'burning water' in the seventh century. The Arab geographer Al-Mas'udi described oil fields in Baku, Azerbaijan, in the tenth century. Arabic and Persian chemists are the first to describe distillation of petroleum in the ninth century, using the distillate for oil lamps.

The modern history of petroleum began in the nineteenth century when James Young, a Scottish chemist, distilled a light thin oil from crude oil collected at Riddings colliery in Derbyshire. Young established the first oil refinery in the world in 1850. The oil was used as a lamp oil, with a thicker oil being used for lubrication. In 1846 the first oil well was drilled in Baku, and the oil industry proper began, with subsequent oil wells being drilled in Pennsylvania and Ontario. The oil was mainly used at that time as a fuel for lamps. Burning it gave off heat and light, and also released the carbon in the form of CO_2, and that's when the trouble started. The carbon that had been trapped in the bodies of the microscopic creatures, which they themselves had got from CO_2 in the air, was returned to the air but in ever increasing amounts.

The oil industry began to grow during World War II, mainly to provide petrol for vehicles. The demand for synthetic materials from petroleum increased, and this was met by replacing costly and sometimes less efficient products with these synthetic materials. This caused petrochemical processing to develop into a major industry. Until the mid-1950s coal was the world's foremost fuel. Coal itself, though, was a problem, as like petroleum it is full of carbon, the origin of which was in plants that had decayed into peat, which had then been compressed into coal. Burning coal, like oil, releases carbon into the atmosphere, also in the form of CO_2. From the 1950s on, oil increasingly replaced coal as the main fuel source, both in the generation of electricity and also in vehicles. We continue to burn billions of tonnes of what we call 'fossil fuel' to fuel our trains, planes and automobiles, provide electricity to heat our houses and keep our computers going. The large rise in fossil-fuel use during the twentieth century is unique in human history because of how it was used. Fertilisers could be made cheaply using oil as the energy source, making food much more plentiful, and this was a major reason for the increase in the human population over the past 100 years. The twentieth century also saw a doubling of the global GDP on four occasions, and all of that can be tracked to fossil-fuel use supporting industry. We are in the thrall of the top three oil-producing countries: Saudi Arabia, Russia and the USA. Incredibly, about 80 per cent of the world's readily accessible reserves of oil are located in the Middle East, with over 60 per cent coming from Saudi Arabia, the United Arab Emirates, Iraq, Qatar and Kuwait.[3] Venezuela has the largest oil reserves of any single country.

So why is the burning of oil or coal problematic for the earth? It's all down to the greenhouse effect.[4] This term refers to a phenomenon whereby a planet's atmosphere warms its surface to a temperature above what it would be without the atmosphere. The term 'greenhouse' in this context was first used by a Swedish meteorologist, Nils Gustaf Ekholm, in 1901. Think of a regular greenhouse: the sun shines in through the glass and warms the air, but the glass prevents a lot of the heat escaping out of the greenhouse, and so the temperature in the greenhouse goes up. Instead of glass, the earth has an atmosphere. Gases in the atmosphere absorb heat from the sun, and then radiate it all around them, including to the surface of the earth, which then warms up. The greenhouse effect was first proposed by Joseph Fourier in 1824, with important evidence for gases in the atmosphere having the capacity to radiate heat being provided by Claude Pouillet, Eunice Newton Foote and Irish scientist John Tyndall. The main greenhouse gases in the atmosphere are water vapour (36–70 per cent), carbon dioxide (9–26 per cent), methane (4–9 per cent) and ozone (3–7 per cent). Tyndall was the first to measure the power of each of these as greenhouse gases.[5] The natural greenhouse effect is, in fact, critical for much of life on earth, as without it our planet would

WOOD ENGRAVING OF A
PENNSYLVANIA OIL WELL FROM 1862.

be too cold and oceans would be frozen solid. The evolution of complex life would not have been possible without the natural greenhouse effect.

The greenhouse effect, however, has become our enemy; because of it, the earth is getting too damn hot. In the late nineteenth century, scientists began arguing that human emissions of greenhouse gases could change the climate. In 1896 Swedish scientist Svante Arrhenius proposed that as humans burned coal or oil, CO_2 levels would rise in the atmosphere, giving rise to what became known as global warming.[6] By the 1930s it was noticed that the US had warmed significantly over the previous 50 years. A lone voice, G.S. Callender, a British engineer, said that there would be further temperature rises because of CO_2. Using US army money for the funding of meteorology research, scientists began to gather data, a firestorm of data, all of which were saying the same thing – the earth's temperature was indeed rapidly rising. This could only be explained by the parallel increase in levels of the greenhouse gas CO_2.

It is now beyond all reasonable doubt that global warming is being caused by greenhouse gases emitted because of human activity.[7] But many are still in denial – and so far President Donald Trump hasn't withdrawn his tweet stating that global warming is a hoax invented by the Chinese to damage US manufacturing. (A bit like his claim that COVID-19 was made in a lab in Wuhan.)

Ice core data is especially important for our assessment of what the climate was like in the past. This is a sample of ice removed from an ice sheet of high mountain glaciers or polar ice caps. The ice builds up year on year, so scientists can take samples and analyse them, and find out the composition of the air for dates way into the past, even as far back as 800,000 years. These data had revealed increasing levels of CO_2 since the early 1800s, when we first began burning fossil fuels.[8] The current level of CO_2 is at its highest for five million years.[9] At that time, the earth's average temperature was 3 °C warmer. Greenland was actually green, and parts of Antarctica had forests. The sea level was 20 metres higher than today, which would mean no Dublin, London, New York, Boston or San Francisco (and a large number of coastal cities). Almost half of the increase in CO_2 has happened since 1990.

Scientists have sounded the alarm.[10] Governments say they are responding, and they have, but nowhere near enough.

THE 45TH PRESIDENT OF THE UNITED STATES.

The measurement of the earth's average surface temperature has shown a rise in temperature of 0.9 °C since the late nineteenth century. The five warmest years on record have occurred since 2010, and current predictions are that the temperature will increase by about 0.2 °C every ten years.[11] The current goal set in Paris at a meeting of the United Nations Framework Convention on Climate Change (UNFCCC) is to limit the temperature increase overall to 1.5 °C. The earth's oceans have shown an increase in temperature of 0.4 °C since 1969. The Greenland and Antarctic ice sheets are melting because of the temperature rise, with Greenland losing an average of 286 billion tonnes of ice per year between 1993 and 2016.[12] The rate of ice loss in Antarctica has tripled in the last decade, resulting in increasing sea levels. Overall, the sea level rose by eight inches in the last century.[13] This brings the threat of flooding as well as a possible disturbance in ocean currents, which depend on salty water. Melting ice caps mean more freshwater. If the

THE PAST FIVE YEARS ARE THE HOTTEST ON RECORD

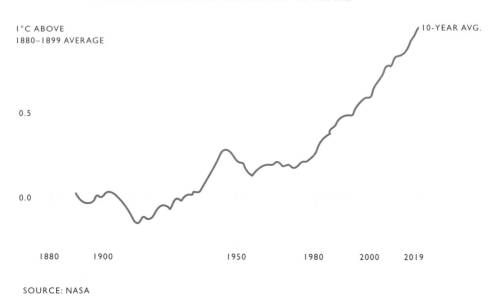

I°C ABOVE
1880–1899 AVERAGE

10-YEAR AVG.

0.5

0.0

1880   1900   1950   1980   2000   2019

SOURCE: NASA

THE WORLD IS GETTING HOTTER EVERY YEAR. DATA FROM NASA CONFIRMS THAT THE LAST 5 YEARS HAVE BEEN THE HOTTEST ON RECORD.

Gulf Stream were to stop flowing towards Ireland, dramatic climate change would happen, such as Ireland becoming much warmer. The increase in CO_2 is also leading to the oceans becoming more acidic, with an increase in acidity of 30 per cent since the industrial revolution, which is very damaging for marine life. This is affecting life in the oceans, most notably the coral reefs, which don't like an acidic environment. Glaciers are retreating, and there is a decrease in overall snow cover, with snows melting earlier and earlier every spring. All of these changes tell us that climate change is happening at an alarming rate, and that the cause is the increase in greenhouse gases in the atmosphere.

The latest report from the Intergovernmental Panel on Climate Change (IPCC) states the situation boldly: 'Atmospheric concentrations of CO_2, methane and nitrous oxide are unprecedented in at least the last 800,000 years. Their effects have been detected throughout the climate system and are extremely likely to have been the dominant cause of the observed warming since the mid-20th century.'[14] Over 99 per cent of scientists have concluded that we are the cause of this problem, with the burning of fossil fuels as the primary reason.[15] This is an extremely high level of consensus, since scientists love to bicker among themselves until the truth is reached. It's not at 100 per cent because some scientists prefer to be mavericks and go against prevailing evidence (or incorrectly fill in the questionnaire). But it is now undeniable. The vast majority of climate change scientists are of the view that the moment of truth has been reached when it comes to humans as the cause of global warming.

The IPCC are especially worried about the oceans. We need our oceans. They provide us with food, 85 per cent of the oxygen we breathe, and regulate the climate. They absorb an amazing 90 per cent of the heat from the warming atmosphere and soak up many gigatons of CO_2. Between 1994 and 2007 they absorbed one third of all the CO_2 from human activity. Without the ocean's surface, temperatures on earth would be 30 °C warmer and there wouldn't be much life. Yet our oceans are under significant threat: since 1970 the global ocean has been getting warmer and from 1993 the rate of warming has more than doubled. More and more fresh water is rushing into the salty oceans, which is bound to have serious consequences, such as the alteration of ocean currents – which will in turn provoke further climate change.

Billions of plankton float through the oceans, harnessing energy from the sun in photosynthesis and so releasing oxygen. They are responsible for half of all the oxygen on earth, the other half coming from land plants. A species called *Prochlorococcus* is especially important for releasing oxygen. There are so many of them in our oceans that they are measured in octillions, which is 10^{27}. They are the lungs of our planet, and yet they too are in serious danger.

The IPCC report also highlights how all of this might affect humans: 680 million people live in coastal communities, which will be inundated by the rising sea levels. A further five million live in the arctic region and 65 million live on small islands that are at risk of submerging. All of these people are at

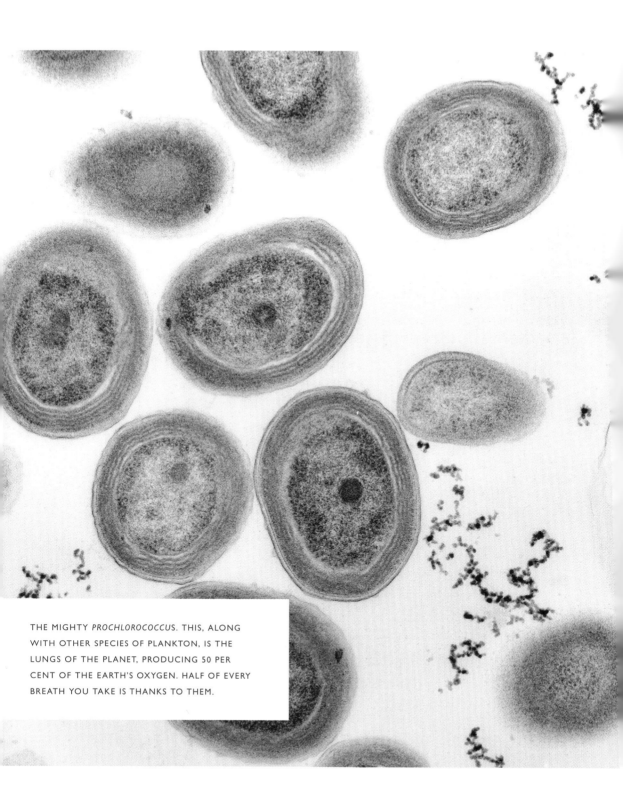

THE MIGHTY *PROCHLOROCOCCUS*. THIS, ALONG WITH OTHER SPECIES OF PLANKTON, IS THE LUNGS OF THE PLANET, PRODUCING 50 PER CENT OF THE EARTH'S OXYGEN. HALF OF EVERY BREATH YOU TAKE IS THANKS TO THEM.

risk of losing their homes, drinking water and livelihoods. Many are in poor communities. The report predicts that the extreme floods, which used to hit once per century, will be yearly occurrences from 2050.

Many species of fish will be unable to survive the warming waters and overall the oceans will become less productive. An incredible three billion of us rely on seafood as our number one source of protein. Mangrove forests and coral reefs are already dying off because of warming waters that are becoming more acidic.

The planet Venus illustrates what can happen when there is a runaway greenhouse effect. CO_2 levels built up in the atmosphere millions of years ago, coming from rocks and soil. The warming of the planet led to more CO_2 being released until eventually the atmosphere was 96 per cent CO_2. This led to a surface temperature of 462 °C and all of the surface water boiling off. The earth is heading in a similar direction. Have we the willpower to stop it? Might we reach what is called a 'tipping point'? This is where climate change starts to accelerate to such a point that it becomes irreversible.

How dire might it all get? The world's climate scientists have a major challenge, and they have to get it right. First they have to build an accurate model of how the earth's climate actually works. Then they have to perturb it, based on likely future scenarios (which mainly relate to human behaviour) and collect the data using supercomputers, which perform billions of mathematical operations per second. Data collection has become much more reliable and efficient, with satellites being put to great use. The importance of this can be seen in the recent finding that Greenland is actually losing three times as much ice than previously thought.[16] A major current concern is melting permafrost, as that would release a vast amount of methane into the atmosphere, accelerating global warming to even faster rates than currently evident from emissions. One major issue is how likely it is that we humans will cut emissions, and thus stop the transfer of carbon from below ground into the air. The transition will need a co-operation between politics, economics and technology. The predictions of what might happen if we don't achieve the Paris goal of reducing global warming to 1.5 °C, are stark.

Trying to slow global warming is a bit like trying to get a supertanker to execute a rapid U-turn. The overall goal is to reverse the 20-fold increase in emissions that occurred over the twentieth century. It will mean replacing everything that burns oil, gas or coal. It will mean recycling or replacing all plastics (which are made from petroleum). It will mean transforming farms all over the world. And all this needs to be done while economies expand to meet the needs of a population, which will be 50 per cent larger than today by 2100. A key challenge is to maintain economic growth: to have what economists call 'green growth'. This may not be possible, as growth has traditionally been linked to increasing emissions since it requires the burning of fossil fuels. Global emissions continue to rise, and actually hit a record in 2018. At the recent UN Climate Summit in New York, 65 countries have committed to reach net zero carbon emissions (taking as much CO_2 out as they are putting in) by 2050. India, for example, has agreed to increase its target for renewable energy fivefold. Only concerted action will work. If one country reduces emissions, but everyone else doesn't, the risks for the planet remain the same. If one country doesn't reduce emissions and all others do, then that country will benefit without putting the effort in.

A recent study found that as few as 100 companies were responsible for 71 per cent of global emissions. Those companies need to be targeted and encouraged to change. At least 650 companies, with a combined total worth of $11 trillion, have signed up to the Science-Based Targets Initiative, which monitors progress on the Paris objectives. These companies agreed to reduce emissions from shipping and buildings.

All of this sounds good, right? It is, but there is still so much to be done. Because the processes that are causing climate change are built into the very basis of the world economy, the measures that will have to be adopted to stop it will also need to be all-encompassing. As an editorial in *The Economist* recently stated, the overhaul might have to involve 'nothing less than a gelding or uprooting of capitalism'.[17] Saving the planet may actually involve 'degrowth': stop aeroplane travel, stop eating meat, ban private cars and divert money from consumption to building infrastructure that is green. Intriguingly, some of these things have happened as a result of COVID-19. Paradoxically, this virus might actually save the planet.

Most importantly, there must be a shift towards sources of renewable energy. Currently 7 per cent of the world's energy comes from the wind and the sun, the two main renewable energy sources.[18] But the cost of such installations has fallen substantially. An increase in renewables may allow for 'green growth', which means economic growth that doesn't harm the planet. The world's biggest offshore wind farm is currently being built in the North Sea. The Hornsea project is constructing 174 wind turbines over an area of 407 square kilometres. It is being made by a British company, and Britain currently has 8 GW of capacity in its offshore wind farms, which is a third more than the next biggest country, Germany, although Germany has nearly three times more wind power capacity. By 2030 Britain will have 30 GW. Wind power is a crucial part of Britain's efforts to hit zero emissions by 2050.

Wind is a perfect complement to solar power, as it blows during the night as well as during the day and is stronger in winter, when there is less sun. Turbines can be placed at sea, avoiding planning issues. The British government has been proactive, funding research into new technologies, and also providing subsidies to companies to encourage development. It recently halted onshore wind development, however, by withdrawing subsidies. All this activity has led to wind power in Britain generating 25 per cent of renewable energy. More impressively, Ireland gets 29 per cent of its electricity from wind power. The International Energy Agency, a Paris-based energy watchdog, is optimistic. In a recent report, it has stated that the falling costs of offshore wind farms will make it competitive with fossil energy within the next decade, forecasting a drop of 40 per cent in cost by 2030. It is also forecasting that by 2040, offshore wind power could be the single biggest source of power generation in Europe. Last year, China installed more offshore and onshore wind farms than any other country.

The move made by the German government to promote the use of solar energy gave rise to great activity in the sector, and prices fell. This might continue. In 2018 world energy demand increased by 3.7 per cent so it's a moving target, but without doubt the use of renewables has to increase. But it turns out that electricity is the easy part. Emissions from electricity-generating plants are less than 40 per cent of all industrial emissions. Industrial processes and transport are a major source of emissions. Only 0.5 per cent of the world's vehicles are electric and this must also increase.

Another approach is negative emissions, or taking CO_2 out of the atmosphere. We already have a great way to do this: plants. They suck up CO_2 and use it in photosynthesis to make more of themselves. We need a lot more plants. And recently scientists at Massachusetts Institute of Technology (MIT) have come up with a new device that can strip CO_2 from the air in a less energy-intensive way than other ways of doing this. A stream of air passes through a stack of charged electrochemical plates, capturing the CO_2, which can then be safely disposed of.

We must also make sure that *Prochlorococcus* and the plankton in our oceans continue to thrive and grow, as they are great carbon-eaters. Apart from getting warmer and more acidic, our oceans are also filling up with waste plastic – approximately eight million pieces of plastic pollution end up in our oceans every day.[19] Every minute, the equivalent of one garbage truck of plastic is dumped into our oceans. Of this, 236,000 tonnes become microplastics – pieces of plastic smaller than your little fingernail. There are enormous plastic patches in the oceans – the one between California and Hawaii is the size of the state of Texas. If we don't stop the plastic from entering the oceans, by 2050 there will be more plastic than fish by weight in our oceans. Many fish we consume, including trout and perch, have microplastics in their bodies. Yet technology might get the tide to turn on plastics in our oceans. The plastic often comes from rivers, which sweep tonnes of waste from the land out to sea. A company called Ichthion have invented a device that sits on the surface of rivers and manages to divert floating objects to the riverbank where a conveyor belt lifts them out and

1. NORTH PACIFIC GYRE   2. INDIAN OCEAN GYRE
3. SOUTH PACIFIC GYRE   4. SOUTH ATLANTIC GYRE   5. NORTH ATLANTIC GYRE

a camera reads them and then diverts anything plastic into waste bins. It can sort up to 80 tonnes of plastic per day, which is sent off for reuse or recycling. Another device can be attached to ships where it can filter plastic particles from the ocean. The battle against plastics in our oceans might yet be won.

THERE ARE ENORMOUS PATCHES OF PLASTIC IN EACH OF THE FIVE MAIN OCEAN GYRES, INCLUDING THE SO-CALLED 'GREAT PACIFIC GARBAGE PATCH' IN THE NORTH PACIFIC GYRE. THIS PATCH IS THE SIZE OF TEXAS.

Scientists are also coming up with 'super corals' by cross-breeding species that are better able to handle warm waters with others that can't, producing heat-tolerant hybrids. This approach is a way of buying time to save corals – ultimately the oceans will have to cool a bit to be sure corals survive. Another important approach is to establish marine reserves to protect species, including fish. Currently, 8 per cent of the world's oceans have some level of protection, with the European Commission claiming that 10.8 per cent of European seas are protected. In 2016 members of the International Union for the Conservation of Nature, which includes 1400 governments and non-government organisations, voted unanimously to protect 30 per cent of the oceans by 2030. This, they feel, combined with the goal of reduced emissions to limit global warming, will give the oceans a chance.

So, there's hope. There has to be hope. And apart from the concerted action of governments it will come from one other place: all of us. Why we continue to wreck the planet is a complicated question, tied into our greed, laziness and the needs of the global economy. What will it take for us to change? We are being told to help the planet by changing our personal choices when it comes to things like diet, travel (especially air travel), energy use in our homes, what we buy in our shops and even the size of our families. To meet the climate change target, we must aim to each produce fewer than three tonnes of CO_2 per year. Currently, the average person in the EU emits 11 tonnes of CO_2 per year. In Ireland, at the height of the last economic boom in the late 1990s, the average was 17 tonnes.[20] Calculations have been done that reveal the level of reduction in carbon emissions per year if we follow some rules. Having one less child will save the parents 58.6 tonnes for each year of a parent's life. Over the course of a year, living without a car will save 2.4 tonnes; one less transatlantic flight will save 1.6 tonnes; adopting a plant-based diet will save 0.8 tonnes; and recycling paper and plastics will save 0.21 tonnes.[21]

Scientists are even measuring the effect different events have on emissions. In a recent study, they examined Oktoberfest in Germany.[22] This is an impressive event by any measure: six million visitors descend on Munich for two weeks of drinking, eating and fun. Roughly 250,000 pork sausages, 500,000 chickens and seven million litres of beer are consumed. Researchers from the Technical University of Munich measured the environmental impact of all that consumption and the results were sobering, to say the least. They sampled the air and found that the festival emitted 1,500 kilograms of methane, an important greenhouse gas (I wonder where that's coming from?). On average, 6.7 micrograms of methane were emitted per square metre per second – ten times that of the city of Boston. But the festival does its best to be green, with a lot of recycling and organic food. Energy from renewable sources is used to provide lighting. And Oktoberfest revellers are encouraged to buy carbon offsets. These all mean that in spite of all the methane, Oktoberfest is likely to survive.

Other than having fewer children, taking local holidays, cycling instead of driving and becoming vegan, what else can we do? Individual purchase of carbon offsets is being touted as a good thing and means giving money to others who will plant trees to offset your CO_2 emissions. This can be done, for example, when you are taking a flight. Various schemes allow you to make a donation to people, often in developing countries, who will then plant trees. Some climate change activists are in favour of citizens participating in groups to advocate collective action, in the form of political lobbying to introduce carbon pricing, meat pricing, ending subsidies for fossil fuel use or making cars much more expensive. These all bring political challenges. The bottom line is we can all help the planet if we choose to: but industry also has to change.

There is also hope in a success story for an environmental policy shift that happened in the 1990s. This began in 1985, when climatologists Joe Farman, Brian Gardiner and Jonathan Shanklin reported a large decrease in ozone levels over the Antarctic stations Halley and Faraday over the previous 30 years. Ozone is a gas that occurs in a layer about 10–50 km above the earth. It is an important filter for potentially damaging UV light coming from the sun. The depletion was worrying because UV light can cause cancer. It was then shown that the depletion in ozone was being caused by chemicals called chlorofluorocarbons (CFCs), which occur in aerosol cans and refrigerants. This led to an international agreement to ban CFCs, and indeed the ozone hole is recovering. This represents a major success story for global environmental policy.

Change can happen. But perhaps the best chance we have of saving the planet is to listen to our children. A remarkable increase in awareness has happened with a new wave of protests occurring all over the world. This can be traced to the young climate-change activist Greta Thunberg. In August 2018 she sat in front of the Swedish parliament building to protest for more

GRETA THUNBERG, SWEDISH ENVIRONMENTAL ACTIVIST, IS HIGHLY CRITICAL OF WORLD LEADERS FOR THEIR FAILURE TO DEAL WITH THE CLIMATE CRISIS.

action on climate change. A month later she announced that she would protest every Friday until the government changed its policy. She called the protest FridaysForFuture, and it has become a global movement. The UK-based Extinction Rebellion is a nonviolent protest group worried about the threat of mass extinction happening because of climate change. The movement's first protest was in October 2018 when 1,500 people gathered in Parliament Square in London. The movement has grown to 150,000 people in 156 different countries. One of the founders of Extinction Rebellion, Roger Hallam, has said he was inspired by the book *Why Civil Resistance Works*.[23] This book gathered data on over 300 violent and nonviolent political campaigns in the past decade and found that nonviolent campaigns were twice as successful as violent ones. The analysis also showed that no regime or leader had remained in power when at least 3.5 per cent of the population had participated in an active protest against them. Isn't it striking that prestigious scientists have said climate change is real and a danger for many years, but the general public and media have generally dismissed or minimised their claims? A little girl saying the same thing has resonated instead. Is this to do with how we rank information content relative to emotional content? As we saw in the chapter on vaccines, people are much more motivated by an emotional message than just information.

Successful movements need some key components. They need to be started by an innovator, someone who gets things going but is persistent and authentic. Greta Thunberg typifies this. Early adopters are then required to build the movement. In the climate movement, this is young people. But sometimes protests can harden attitudes. If you are sceptical about climate change, then having your route to work disrupted by climate activists might make you even more sceptical, or antagonistic, as happened when Extinction Rebellion disrupted the Tube system in London. There is also an optimum window in which movements have an effect, which is three years on average. What Extinction Rebellion or FridaysForFuture have going for them is the youth face of the movement. Lots of new and different people are getting involved, in what is seen as the most pressing issue for the current generation. Let's hope for all our sakes it has an impact. With the help of political leadership we can turn this supertanker around before it's too late. If we don't, the world will be a totally different place in 50 years' time, and all because of our inability to act on the compelling science that has been done on this most important of topics: the health of the planet. We caught a glimpse of what a clean, healthy world

THIS PHOTOGRAPH, NAMED 'EARTHRISE', WAS TAKEN IN 1968 BY THE CREW OF THE APOLLO 8 MISSION. IT REVEALED THE BEAUTY AND FRAGILITY OF THE EARTH, AND IS WIDELY CONSIDERED TO BE THE MOST IMPORTANT ENVIRONMENTAL PHOTOGRAPH EVER TAKEN.

might look like during the COVID-19 pandemic. The huge decrease in human activity led to the air clearing over many countries, with levels of CO_2 and nitrogen dioxide plummeting. The greenhouse effect can then be lessened, and there can be cleaner air, since nitrogen dioxide is a significant pollutant that damages lungs. It happened within a month or so, telling us we can act if we need to, and the response can be rapid.

The bottom line: global warming is being caused by human activity. This has to change. We need to act now, and fast. As the children protesting have shown on placards and posters: *There is no Planet B.*

CHAPTER 14

WHY SHOULDN'T YOU LET PEOPLE DIE IF THEY WANT TO?

—

'Euthanasia – that sounds good
A neutral Alpine neighbourhood
Then back to Britain all dressed in wood
Things were going to get worse – apparently'

—

John Cooper Clarke, 'Bed Blocker Blues'

MY FATHER ASKED me to kill him. He was 74 years of age, had suffered a stroke at 71 and couldn't speak properly. He also had paralysis on his left side. He was a widower and was in a nursing home. We had tried home helps who came in every day, but he was also severely depressed, which made things difficult. He would say to me in his slurred voice: 'You work in a lab. You have the chemicals to do it.' He also regularly said, 'If I were a horse, you'd shoot me.' My father had a dark sense of humour that I loved, so I used to brush these conversations off as gallows humour, but I knew he meant it. I would sometimes cry when I left him in his room, in his own separate hell.

Should I, out of sympathy and love, have bumped him off? That would have been murder. But what if the law had allowed me to help him die? How would that have worked and would I have had the guts to do it? So, let's look at euthanasia: how it's done and what the safeguards are. Will a time come when it will be as routine as childbirth, as we head towards a population where the majority are sick and old, with lots of older people actually wanting to die? Or will medical advances, both in the discovery of new treatments for diseases and better palliative care, make euthanasia unnecessary? As scientists, we must shirk from nothing and face this topic head-on, from as scientific a point of view as possible.

There are plenty of stories in science fiction featuring euthanasia. Most famously, in the 1976 film *Logan's Run*, a dystopian world is described in the year 2116 (which isn't that far away). In order to maintain equilibrium in the consumption of resources, when people reach the age of 21, they must die by euthanasia. Their 21st birthday is known as 'Lastday'. Instead of getting the key to the door, they report to the 'Sleepshop' and are given a pleasure-inducing toxic gas. Their age is revealed by a crystal in the palm of their right hand. This crystal changes colour every seven years, blinking red and black on Lastday – before finally settling on black. Our world won't become like *Logan's Run*, but the prospect of euthanasia becoming common is not as unlikely as it once would have seemed.

LOGAN'S RUN, A 1976 MOVIE SET IN THE 23RD CENTURY, DEPICTED A SOCIETY WHERE EVERYONE IS KILLED AT THE AGE OF 30.

Euthanasia comes from the Greek for 'good death'. There are two types: in the first type, active euthanasia, one person performs the act of euthanasia on another, who has given their consent; the second type is called 'assisted suicide', in which everything up to the last step is provided by the other person.[1] The distinction is important, with assisted suicide being defined as 'intentionally helping a person die by suicide by providing drugs for self-administration, at that person's voluntary and competent request'. Effectively, euthanasia is suicide by someone who is too infirm to kill themselves without someone else's help. Somewhat pointlessly, suicide remains a crime in some countries, which has consequences for things like the fate of the dead person's estate. The British House of Lords Select Committee on Medical Ethics defines euthanasia as 'a deliberate intervention undertaken with the express intention of ending a life to relieve intractable suffering'. Yet in the Netherlands and Belgium it is defined slightly differently as 'termination of life by a doctor at the request of the patient'. This means that it doesn't necessarily have to involve the relief of suffering, which is an important distinction. The medical understanding of what suffering is can be hard to pin down. Does psychological suffering count, and how would that be measured? Perhaps the Dutch and Belgians have simplified the definition for that reason.

Active euthanasia is legal in Belgium, the Netherlands, Luxembourg, Colombia and Canada. Assisted suicide is legal in Switzerland, Germany, the Netherlands, the State of Victoria in Australia and the US states of California, Oregon, Washington, Montana, Washington DC, Colorado, Hawaii, Maine, Vermont and New Jersey.[2] It is illegal in all other countries, as is non-voluntary euthanasia (where the patient is unable to give consent). Although legal in the countries mentioned above, it is only allowed under certain circumstances and requires the approval of two doctors and in some places a counsellor. Treatment or medical support being withdrawn because it is considered futile will also hasten death but is not illegal. Ethically, what distinguishes euthanasia from murder is intentionality. The intention of the person who is performing euthanasia is to relieve suffering in as painless a way as possible for a person who has given their consent. It all seems so reasonable, doesn't it? And countries like the Netherlands have always been reasonable when it comes to how humans should and shouldn't behave, treating adults like adults (as we saw in Chapter 7).

What about Ireland? Three legal cases are informative about where Ireland might be going. In 1995 the Supreme Court gave permission for a feeding tube to be removed from a woman who was in a persistent vegetative state for over 20 years so that she could die a natural death. The court, however, stressed that it would not condone any attempt to end a person's life through positive action.[3]

In another case, Marie Fleming, a lecturer in University College Dublin, was in the final stages of multiple sclerosis.[4] She and her husband mounted a legal challenge against the state. Marie could no longer use her limbs and said she wanted to die at a time of her own choosing with the assistance of her partner since she was not physically able to take direct action herself. They lost their case. The court ruled that the Constitution does not contain either a right to suicide or to arrange for the end of one's life. The case attracted great attention, and Marie's courage drew a lot of admiration. During the court case, Marie said, 'I've come to court today, while I can still use my speech, to ask you to assist me in having a peaceful, dignified death in the arms of Tom and my children.' Marie was challenging the absolute ban on assisted suicide in the Criminal Law (Suicide) Act 1993. She made the case that the law disproportionately infringed her personal autonomy rights under both the Irish Constitution and the European Convention on Human Rights. Marie subsequently died a year later of natural causes.

In 2013 a woman called Gail O'Rourke was charged with assisting the suicide of her friend Bernadette Forde between 10 March and 6 June 2011.[5] Bernadette died in 2011 after taking a lethal dose of barbiturates ordered by Gail on her behalf from Mexico. There were three charges against Gail: ordering the drug which Bernadette would take to kill herself; arranging Bernadette's funeral ahead of time; and planning a trip to Zurich, where

Bernadette hoped to die. The plan was prevented when the travel agent informed the Garda. Gail O'Rourke was acquitted in 2015 of three counts of assisting the suicide of Bernadette Forde.

The current situation in Ireland is that both euthanasia and assisted suicide are illegal under Irish law. Euthanasia is regarded as either manslaughter or murder. The Health Services Executive provides a list of alternatives to euthanasia in Ireland.[6] First, a patient can refuse treatment. If a person knows that their capacity to consent may be affected in the future, they can pre-arrange a legally binding advance decision (called a living will) that sets out the treatments that they do not consent to. If a patient is undergoing surgery that could cause respiratory or cardiac arrest, they have the option of making it clear that they do not want to be treated with cardiopulmonary resuscitation. The legal basis for this is not clear in Ireland, but most doctors are likely to respect it. This is known as 'do not resuscitate', or DNR. This is allowed because of the low success rate of resuscitation, and the likelihood of serious complications, including brain damage, and is usually allowed for patients that have a terminal illness. A second alternative to euthanasia is palliative sedation. This involves giving a person medication that will make them unconscious, unaware of the pain and ultimately hasten their demise by affecting their breathing. It carries the risk of shortening life but is widely used. Third, doctors can withdraw life-sustaining supports if it is clear that the prospects of the patient recovering are nil. Patients are usually heavily sedated when life support is withdrawn, allowing them to die peacefully.

The debate around euthanasia in Ireland continues with no apparent change on the horizon. The discussion on euthanasia began in the mid-1800s when morphine began to be used to 'ease the pains of death'. Anna Hall is famous as an early strong advocate for euthanasia in the USA.[7] She had watched her mother die after a long battle with liver cancer and dedicated her life to ensuring that others would not have to endure the suffering of her mother. In 1906 she pressed for legislation in Ohio but was unsuccessful. The UK also had strong proponents for euthanasia, with The Voluntary Euthanasia Legali-

KING GEORGE V (1865–1936) WAS GIVEN A FATAL DOSE OF MORPHINE AND COCAINE, WHICH HASTENED HIS DEATH FROM CARDIO-RESPIRATORY FAILURE.

sation Society being founded in 1935 by Charles Killick Millard – it's now known as Dignity in Dying. An early case of euthanasia happened in the UK when King George V was given a fatal dose of morphine and cocaine to hasten his demise from cardio-respiratory failure, although this wasn't made public until 50 years later.[8] That it happened suggests that the practice of euthanasia might not have been a rare event in the UK. In 1949 the Euthanasia Society of America sent a petition to the New York State Legislature requesting that euthanasia be legalised. It was signed by 379 leading Protestant and Jewish ministers.[9] A similar petition had been sent in 1947, signed by over 1,000 New York doctors, but no legal changes happened.

Since then euthanasia has been debated from time to time, and it is highly likely that the question will come up more and more as the population continues to age. The debate centres on four issues: the right of people to choose their fate; that helping someone to die is better than leaving them to suffer; that the ethical difference between the commonly practised 'pulling of the plug' and active euthanasia is not substantive; and that permitting euthanasia will not necessarily lead to unacceptable consequences. This is certainly the case in countries like the Netherlands and Belgium, where euthanasia has mainly been unproblematic (although, as we will see, problems might be emerging).

One of the more common issues that arises during debates on legalising euthanasia is problems around consent. Perhaps the person is not competent to make the decision (determining competence is not straightforward). Perhaps the person feels that they are a burden on medical services or on their family. How do we know unscrupulous friends or relatives aren't pressing the person towards it? Do hospital personnel have an economic incentive to encourage consent? Shouldn't better palliative care make euthanasia unnecessary? And what about medical advances that are making previously incurable diseases now potentially curable? A good example of this is melanoma, a type of cancer that is now curable in some cases because of a process called checkpoint blockade.[10]

Religious views on euthanasia vary.[11] The Catholic church condemns it as morally wrong, as do several Protestant churches, including the Episcopal, Baptist, Methodist and Presbyterian churches. The Church of England accepts passive euthanasia but is against active euthanasia. Islam opposes the taking of life whatever the reason, while in Judaism it remains unacceptable, although hotly debated.

Euthanasia is an important topic for ethicists. It raises several tortuous moral dilemmas, including whether there is a moral difference between killing someone and letting them die. The core of the ethical issue is different ideas on the meaning and value of human existence. Apart from ethical considerations, the main thing holding back legislation to allow for euthanasia may be one of squeamishness as a result of the complexity of the issues and fear of attack from voters. Some countries may simply not want to legislate for something which feels wrong or is unpleasant, or the issues raised during the debate are difficult to resolve. And so euthanasia remains illegal in the majority of countries.

But there is a growing acceptance of euthanasia in the general public. Numerous surveys have been done in many countries and opinion in favour of assisted suicide appears to be on the increase. In 2013 a massive survey (scientists like large surveys because they are likely to give a more accurate picture) was carried out in 74 countries.[12] Overall, 65 per cent voted against physician-assisted suicide but in 11 of the 74 countries, the vote was mostly in favour. In 2017 a Gallup poll found that 73 per cent of US participants were in favour – a clear majority.[13] Fifty-five per cent of weekly churchgoers were in favour, while the number was 87 per cent in favour for non-churchgoers. In a 2019 survey of 2,500 people[14] in the UK, more than 90 per cent believed that assisted euthanasia should be legalised for those suffering from a terminal illness. Eighty-eight per cent believed that it was acceptable for people living with dementia to receive help to end their lives, provided that they consented before losing their mental capacity. Such a high level of support is likely to put pressure on politicians to legislate. In another survey in the UK, 52 per cent of people said they would feel more positive towards their MP if they supported assisted dying, compared to just 6 per cent who said they would feel more negative.[15] In Ireland, a recent poll revealed that 63 per cent of the population was in favour of euthanasia – not that dissimilar to the 64.5 per cent who voted to change legislation on abortion.[16] Younger people were less likely to be supportive: 48 per cent of 18–24-year-olds were supportive, contrasting with 67 per cent of 35–44-year-olds. For people aged over 55, support dropped to 49 per cent.

So what concerns people, apart from reasons to do with religious belief? Guidelines and safeguards are important.[17] Physicians and counsellors are all involved in assessing people requesting euthanasia. These vary in the countries where euthanasia is practised. In the US, Canada and Luxembourg, the person must be over 18. In the Netherlands, the age is 12, while in Belgium there is no age limit as long as the person has the capacity for discernment. As regards safeguards, these also differ. In the US, there is no need for unbearable pain or any symptoms. In the Netherlands, Belgium and Luxembourg, patients must have 'unbearable physical or mental suffering' with no likelihood of improvement, although the person doesn't have to be terminally ill. One issue that may emerge is that there is a danger that people with severe long-standing depression might want to have their life ended if they are terminally ill. This might be difficult to evaluate, as many with a terminal illness may also be clinically depressed. There are also differences when it comes to procedural requirements. In the US, assisted suicide must involve a 15-day period between two oral requests, and a 48-hour waiting period after a final written request. In Canada, there is a ten-day waiting period after a written request. In Belgium, there is a one-month waiting period. The Netherlands and Luxembourg do not have any waiting period. Studies have shown that across all places where it is legal, around 75 per cent of people who undergo assisted suicide are suffering from terminal cancer.[18] The next-highest condition on the list is amyotrophic lateral sclerosis (also called motor neurone disease) at 10–15 per cent. Pain is not that common as a motivating factor, with issues such as loss of autonomy and dignity being more important.

How is it done? Euthanasia is performed by a doctor administering a fatal dose of a suitable drug to the patient. Assisted suicide involves the doctor supplying the person with the drug to self-administer it. The group of drugs most commonly used is barbiturates,[19] which work by causing the brain and nervous system to slow down. This causes the respiratory system to fail, leading to death, with the person fully sedated. The same drugs used in small doses are used to treat insomnia, but the dose is so high in euthanasia that you never wake up. Barbiturates target GABA receptors by acting like GABA, the primary inhibitory neurotransmitter in the brain.[20] When this receptor is turned on by GABA or barbiturates, it moves chloride through the membrane of neurons, which dampens their activity. Barbiturates bind to several pockets within the GABA receptor, which are different from where GABA itself binds. They also bind to and inhibit receptors for excitatory neurotransmitters, including the AMPA and kainite receptors. In short, they have a double-whammy effect: mimicking the inhibitory neurotransmitter (GABA) and blocking excitatory neurotransmitters. It's a bit like applying the brakes on a car while releasing the accelerator. As a result, the brain starts to slow down, and this leads to respiratory failure.

The name 'barbiturate' was coined by the German chemist Adolf von Baeyer, who made the first one (barbituric acid) in 1864;[21] Baeyer and his colleagues went to a local tavern to celebrate their discovery, where there was a celebration going on for the feast of St Barbara. Another story says that Baeyer synthesised barbituric acid from the urine of a Munich waitress called Barbara. It took until 1903 for barbituric acid to find a use, when it was found to be effective at putting dogs to sleep. During World War II, soldiers in the Pacific region were given barbiturates (nicknamed 'goofballs') to allow them to tolerate the heat and humidity, as even at low doses barbiturates reduce the respiration rate, making working in the heat less stressful for the lungs and heart. Many soldiers returned with a lifelong addiction, made worse

GERMAN CHEMIST ADOLF VON
BAEYER (1835–1917) MADE THE FIRST
BARBITURATE IN 1864. THE NAME WAS
TAKEN FROM ST BARBARA, WHOSE FEAST
DAY FELL ON THE DAY THE DRUG WAS
DISCOVERED. ANOTHER STORY IS THAT
HE SYNTHESISED IT FROM THE URINE OF
A MUNICH WAITRESS CALLED BARBARA.
HE MIGHT NOT LOOK IT, BUT ADOLF
CERTAINLY KNEW HOW TO PARTY.

by doctors continuing to prescribe barbiturates. In the 1950s and 1960s, barbiturates were prescribed for anxiety and insomnia, but because of their addictive nature, they were increasingly replaced with another drug type called benzodiazepines, which include Valium, the tradename for diazepam. Marilyn Monroe, Brian Epstein and Judy Garland all died of a barbiturate overdose.

The main types of barbiturates that are used for euthanasia are secobarbital and pentobarbital. Pentobarbital is also used in the execution of convicted criminals in the USA. These can be used alone or in combination. They are safe (in the sense that there are no obvious side effects) and cause a peaceful, swift and uneventful death.

So how likely is it that the rest of the world will follow the likes of the Netherlands and Belgium and relax the laws around euthanasia?

Baby boomers who campaigned so vigorously for contraception and abortion have become old, with debilitating illnesses. Will they now campaign for their own deaths? What will our society be like if we become like the Netherlands, where almost everyone knows someone who has died by euthanasia? Some doctors there are starting to worry that things have gone too far.

A journalist named Christopher de Bellaigue recently reported on how a doctor in the Netherlands, Bert Keizer, was called to the house of a man dying of lung cancer.[22] The man had felt his time had come. Keizer arrived with a nurse to assist him, and they found 35 people around the bed, drinking, laughing and crying. The man cried out 'OK guys!' and everyone fell silent. Small children were taken from the room and the doctor administered the lethal injection. This is apparently a typical scenario. Dr Keizer works for the End of Life Clinic in the Netherlands, which in 2017 performed euthanasia on 750 people out of a total of 6,600 in the country as a whole. He is of the view that euthanasia is much better than regular suicide, which leaves deep wounds in loved ones left behind. In 2017, 1,900 Dutch people committed suicide – but a further 32,000 died under palliative sedation. The future may well look like the Netherlands.

But the situation in the Netherlands is also raising concern. Where do you draw the line? This was always a concern in debates on euthanasia. The idea that allowing it is a slippery slope, where a measure which aims to provide relief from suffering for cancer patients is expanded to include people who might otherwise live for many years. In the Netherlands, an ethicist called Theo Boer was given the task of reviewing every act of euthanasia between 2005 and 2014.[23] He is openly critical of euthanasia in the Netherlands, especially since the law changed in 2007 to include a range of conditions, while the term 'unbearable suffering' as a reason for euthanasia was loosened. Many Dutch people now legally state that they are to be euthanised if their mental state deteriorates beyond a certain point – say, unable to recognise relatives. This has given rise to euthanasia in dementia patients, and some are uneasy at this development.

Medical ethicist Berna Van Baarsen resigned as a euthanasia review board member because of the growing numbers of dementia sufferers who were being euthanised based on a prior instruction. He resigned because of one horrible case. A patient who had instructed that she should be euthanised prior to dementia resisted when the time came (as judged by her having advanced dementia) and had to be restrained by her family while the doctor adminis- tered the lethal injection.[24] The issue of conditions which aren't terminal being the reason for euthanasia is currently being debated in the Netherlands, yet it's unlikely that the law will change to prevent euthanasia for dementia. In other countries that are considering making euthanasia legal, perhaps the Dutch experience will lead them to restrict it to terminal patients only.

Public and legal opinions might be shifting among the Dutch. Recently, a doctor was tried for failing to verify consent before performing euthanasia on a patient with dementia.[25] The 74-year-old patient, who died in 2016, had previously stated in writing that she wanted to be euthanised. The judges ruled that the doctor was acting on the woman's instructions, but prosecutors argued that the doctor had failed to ensure the consent of the woman, who might have changed her mind. They said a more intensive discussion should have taken place. The trial is considered an important test case, as there are likely to be more cases of patients suffering from diseases such as Alzheimer's disease who have requested euthanasia when fully *compos mentis*. The issue came down to whether someone who makes a choice when they are of sound mind should be held to that choice when they no longer are. The judges in the case said that they should be held to that choice, and there was a small round of applause in the courtroom when the verdict was read out.

CHRISTIAN DE DUVE (1917–2013) WON THE NOBEL PRIZE IN 1974 FOR THE DISCOVERY OF THE LYSOSOME, A KEY PART OF EVERY CELL. AT THE AGE OF 95 HE WAS DIAGNOSED WITH TERMINAL CANCER AND DECIDED TO DIE BY EUTHANASIA.

The question now becomes: at what point do you stop checking if someone wants to die, and should they still have the power to do so even when they are no longer in control of their mental functions?

When I think about the rights and wrongs of euthanasia, I think about two people. First, Christian de Duve, a famous Belgian biochemist[26] who won the Nobel Prize in 1974 for the discovery of the lysosome, a tiny sack full of enzymes that is inside all cells. The lysosome is the garbage disposal system for cells: it destroys parts of the cell that are old or worn out and can digest a cell whole when it becomes old or damaged. Lysosomes are a bit like a euthanasia machine for the cell. I had the pleasure of hosting de Duve at a conference that commemorated the 50th anniversary of the 'What is Life?' lectures given by Erwin Schrödinger in Trinity

College Dublin in 1943. These lectures had sparked a revolution in biology that gave rise to many advances. Christian died by euthanasia in Belgium at the age of 95. He had been suffering from a number of ailments, including terminal cancer. A friend and colleague, Günter Blobel, said de Duve wanted to make the decision while he was still able to do it and not to be a burden on his family. Christian spent the last month of his life writing to friends and colleagues to tell them of his decision to end his life. In an interview with the Belgian newspaper *Le Soir*, published after his death, he said he intended to put off his death until his four children could be with him.[27] He also said he was at peace with his decision, saying, 'It would be an exaggeration to say I'm not afraid of death, but I'm not afraid of what comes after, because I'm not a believer.'

The bottom line is that euthanasia, when properly regulated, can give us hope of a better quality of death. We must also strive for scientific advances in bringing better treatments or palliative care for those who suffer.

I will leave you by telling you about the second person who is on my mind when I think of this topic. I go back to my father. During the winter of 1995–6, Dad suffered several bouts of pneumonia, almost dying on one occasion. In January of 1996 his GP asked to see me. He suggested that perhaps he wouldn't prescribe another course of antibiotics and would see if my Dad could fight the latest bout on his own. I knew what he was saying by the way he looked at me. My dad died peacefully of pneumonia (or 'the old man's friend' as he used to call it) in his sleep on 20 February 1996, with me sitting beside his bed, holding his hand.

Not a bad way to go, Dad.

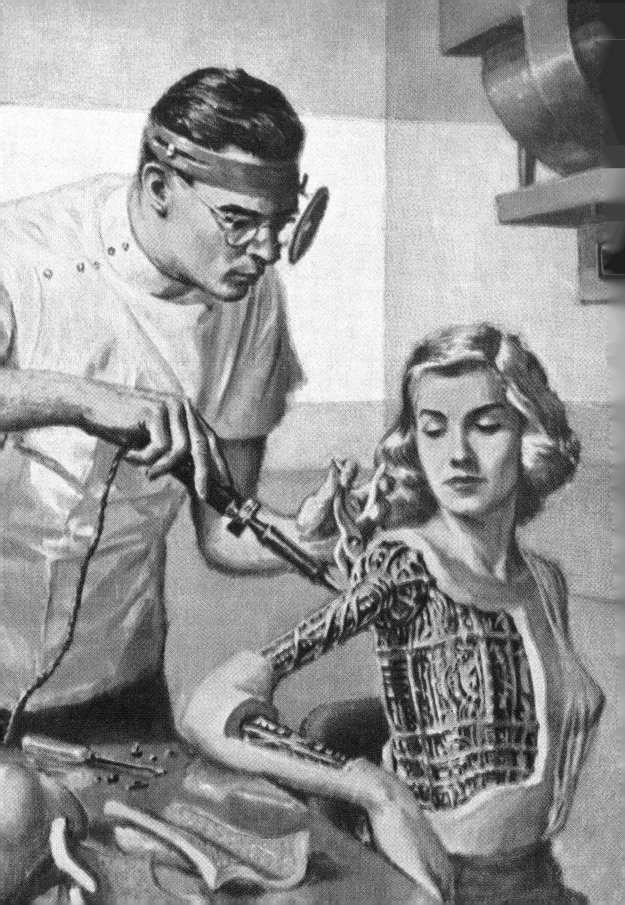

CHAPTER 15

WHAT HAVE YOU GOT TO LOOK FORWARD TO?

—

'It's tough to make predictions,
especially about the future.'

—

Yogi Berra, baseball player

WHEN I WAS a boy, I really wanted a jetpack. I'd seen one in a cartoon called *The Jetsons* where Elroy spent most of his time floating around in one. I used to imagine jetpacking my way to school. I really believed that I might get one in the future. My father (who was born in 1921) used to tell me how when he was a boy there were no jet planes and, unimaginably for me, he grew up with no television. I used to think that when I grew up and had kids, I would tell them how unbelievably in my own childhood there were no jetpacks – and no mobile phones. I have told them about that last one and they still refuse to believe me. Sadly, jetpacks have yet to become commonplace.

But most of all I thought maybe the future would be like *Star Trek*. The writers tried hard to imagine what technology might look like for each series. One of the most interesting episodes, at least from an Irish perspective, was 'High Ground', broadcast in 1990. A crew member is taken hostage by rebels on the planet Rutia IV. One of her colleagues draws an analogy with a conflict on earth several centuries earlier – the Troubles in Northern Ireland – making the comment that they were ultimately resolved by 'the Irish unification of 2024'. Did they get it right? Might Brexit lead to a United Ireland? *Star Trek* managed to get several things right about technologies we

STAR TREK: THE NEXT GENERATION

now have. Futurologists are always trying to predict the future, and science fiction has a part of play in that enterprise. What might our future look like?

I'd love to move through time to the future and see how many of the things I've covered here will have become true because of science. Wouldn't it be great if all my bottom lines came true? As yet, a time machine hasn't been invented, although according to some physicists that is not completely beyond the bounds of possibility.[1] Will vaccination be made mandatory and will we defeat most infectious diseases? What will our addiction to technology look like? Will all drugs be legalised for adults? Will euthanasia be widely available, with clear regulation? Will there be no more bullshit jobs as automation becomes the norm, and major changes happen in the way we work and the kinds of jobs we do? Will the idea of men and women as different types of human be superseded by a more complex and nuanced view of humanity? Will obesity no longer exist (because of medical intervention or better nutrition)? Will many diseases (including mental health conditions, addiction and cancer) be distant memories, like TB and polio are now, in the developed world at least? Will we save the planet? If all those things come to pass, won't the world be a much better place for all humans and shouldn't we aim to achieve each of them?

It's worth looking at *Star Trek*, and indeed other science fiction stories such as *Black Mirror,* a bit more closely, to see if we can save the world. Instead of being predictive of the future, they are sometimes commentaries on the present. If we disregard aliens and spaceships, a lot of science fiction is about the pressing concerns of today. The impact of artificial intelligence. The dangers of the ecological crisis. How power can be misused in totalitarian regimes that control technology. Another common theme in science fiction is changing attitudes to gender politics – a world in which gender doesn't matter and can be chosen or modified at will. What science fiction can help us with is a guide to the future, given what is happening now.

Perhaps somewhat surprisingly, science fiction can help governments and corporations plan for the future. The French government have set up the Defence Innovation Agency,[2] which has a team of science fiction writers proposing scenarios for the future. The engineering firm Arup have commissioned writers to describe four scenarios that might happen as a result of climate change.[3] Google, Microsoft and Apple all employ science-fiction writers as consultants. For businesses, science fiction can free up the mind from constraints. Science fiction can inspire people who work in the technology sector to come up with new products and services. Motorola have stated it was the hand-held wireless communicators used in *Star Trek* that motivated them to make the first mobile phone.[4] Amazon's Alexa voice assistant was inspired by the talking computer on the *Starship Enterprise*.[5] The Kindle was inspired by an electronic book that featured in a novel by Neal Stephenson.[6] Instagram CEO Adam Mosseri has said that an episode of the science fiction series *Black Mirror* inspired him to hide public-facing likes on Instagram posts.[7] An episode called 'Nosedive' has a storyline involving a world where people rate one another on a scale from 1 to 5 based on their interactions with one another. Your score influences how people in society treat one. In the episode, the main character spends most of her time obsessing with improving her rating to improve her real-world opportunities with disastrous consequences for her mental health. The technology innovators of the future would do well to read the science fiction of today.

The politics of *Star Trek* is also interesting. Its main creator, Gene Roddenberry, had fought in World War II, and at the time he created *Star Trek*, he was horrified that there might be another war, this time with Russia. Several episodes involve totalitarian regimes, sometimes run by computers, which rob people of their liberty. This is not too far from the situation today, where Russia has been accused of influencing the US presidential election and the Brexit vote of 2016 (see Chapter 1) by using computer technology. In *Star Trek*, planets joined together into a United Federation in the year 2161. The Federation has a key technology – warp drive – described as 'superluminal spacecraft propulsion'. It allows spacecraft to travel at speeds many orders of magnitude faster than light. Space is warped to allow this to happen.

Scientist Zefram Cochrane discovered the space warp in 2063. The first test of warp drive led to first contact with aliens, the Vulcans.

The United Federation of Planets is headquartered in San Francisco. Roddenberry had NATO in mind when he described it, with the Klingons representing the Soviet Union. It presents an optimistic view of the future. The economy in the Federation is described as 'post-scarcity' and has evolved beyond government-controlled monetary systems. Money is obsolete because of replicator technology. Many goods and products are easy to make with replicators. The Federation has a president, a cabinet and a supreme court. The Federation's military/exploration arm is Starfleet Command. The council is made up of delegates from member sovereignties. In 2267 Captain Kirk says there are a thousand planets in the Federation, which continues to expand.

The two key technologies that define the Federation are warp drive and replicators. How likely is it they will come true? The only means of space transport that we now have is rocket propulsion, which hasn't changed much since the 1960s. This means space travel is constrained by chemistry. Burning combustible fuel with storable or cryogenic oxidisers remains the only way to boost rockets. In 2018 the US Defence Intelligence Agency made public a 2010 report, which described possible approaches to travel faster than light.[8] It was met with scepticism. There may be other options: nuclear-powered engines might be used. Space infrastructure has also got much better. The international space station has been called 'Starfleet in gestation'. The next step is the lunar gateway, which is planned for 2026.[9] It will orbit the moon and allow astronauts to return to the moon's surface.

Whatever about warp speed, a lot is happening in the aviation industry – but if a time traveller from 1968 visited an airport today they wouldn't notice much difference in the aeroplanes, although they would notice a huge difference in social distancing as a result of COVID-19. The 1970s looked promising with the advent of supersonic jets like Concorde, but they proved to be uneconomic. If the aviation industry was a country, it would rank among the worst countries in the world for carbon dioxide emissions.

These have risen by 70 per cent since 2005, with a further seven-fold increase predicted by 2050.[10] There are currently around 200,000 pilots and this is predicted to increase to 600,000, unless flying is restricted because of the damage it is doing to the planet or because of the constraints imposed by COVID-19.[11] It's likely that pilotless aeroplanes will become more of a reality as artificial intelligence is used to fly aeroplanes, although a fully pilotless aeroplane isn't envisaged yet. A frequent flyer tax might be imposed but this would discriminate against poorer countries (where the growth in flying is happening). Flight rationing might be an option, where people will have a limit on kilometres they can fly per year. High-speed rail lines might replace flying. Electric aeroplanes are being developed, with more powerful batteries and lighter, more powerful engines, which are less harmful to the environment.[12] Electric flying taxis are being developed, which may eventually lead to a future often seen in science-fiction films – flying cars in the skies of cities. Solar-powered airships are also being developed – and may usher in an era of slow travel. An airship would take 44 hours to cross the Atlantic.

One of the more outlandish ideas is to build an orbital ring around the earth.[13] This would comprise a strong steel cable about 80 km above the earth. It rotates, creating a force. Two magnetic levitation (Maglev) train tracks on the underside of the ring would transport passengers at

THIS IS WHAT THE DART WILL LOOK LIKE IN 2050.

ELON MUSK, SOUTH AFRICAN ENTREPRENEUR. HE FOUNDED TESLA, AN ELECTRIC VEHICLE COMPANY, AND SPACEX, THE FIRST PRIVATE COMPANY TO TAKE ASTRONAUTS TO THE INTERNATIONAL SPACE STATION.

incredible speeds. It is currently envisaged that a Maglev on an orbital ring would allow passengers to get to Australia from Europe in 45 minutes.[14] Who knows, COVID-19 might well hasten its development, as we might no longer be able to take long-haul flights.

New supersonic aeroplanes are also being developed. There's no doubt that when it comes to speeds, aeroplanes have actually gone backwards since the heyday of Concorde. New companies are emerging to bring supersonic flight back. Boom Supersonic is developing a commercial aeroplane that will fly at over twice the speed of sound, or Mach 2.2, with lower running costs than Concorde.[15] The Space Liner is being developed by the German Aerospace Center (DLR) and is predicted to fly at 25 times the speed of sound.[16] This will travel on the edge of space and get you from London to Australia in 90 minutes, much faster than the current best travel time from London to Perth, which takes 16 hours and 35 minutes. I took that flight. It was horrendous.

Elon Musk's company, SpaceX, is at the forefront of developing spacecraft.[17] It became the first to deliver cargo to the International Space Station in 2012. He has built the *Crew Dragon*, with help from NASA, which has flown astronauts to the space station. He has also built a large number of communications satellites, and has launched 122 of these in a constellation called Starlink.[18] In total, SpaceX says that 12,000 satellites will be deployed by the mid-2020s. Musk hopes that Starlink will revolutionise broadband communications, providing it to underserved areas on earth and beyond into space, including to Mars. He is also co-funding with Microsoft (who provided $1bn) a major initiative in artificial intelligence called OpenAI and is helping the establishment of a base on the moon, called 'Moon Base Alpha'.[19] His main ambition, though, is to get a spaceship to Mars. NASA have the aim to get humans to Mars by 2030. Musk feels that humans must become an interplanetary species to combat the threat of asteroids and the risk of human-created catastrophes on earth, notably nuclear war or engineered viruses, the latter becoming all the more relevant since COVID-19. His company recently announced their new starship, *SN1*, which will start test flights soon. It is 165 feet tall with a giant rocket called *Super Heavy*. A Japanese billionaire has booked an around-the-moon voyage on the craft in 2023 and is already looking for a female life partner to go with him.[20] Any takers?

Although we may be a long way from developing warp speed, we may be closer to the replicators that were used in *Star Trek*. 3D printing technologies are proceeding apace. This is described as additive production. Traditional production involves cutting away materials to make the final product. 3D printing adds layers of material to build the object. It is being used to make medical prosthetics, food including chocolate, crackers, pasta and fibrous plant-based meat, clothes including shoes and dresses, and even components for cars, planes and boats. 3D printing is likely to advance even further, with proposals to print whole houses and their contents. Who knows where it might lead?

Apart from warp speed and replicators, the other striking thing about *Star Trek* is medical procedures. When Dr McCoy visits a hospital in the 1980s by travelling back in time in *Star Trek IV: The Voyage Home*, he compares the medical procedures he sees to the Dark Ages. Geordi La Forge wears a visor that allows him to see, despite being blind from birth. A device somewhat similar to the visor has been invented. In 2005 a team at Stanford University used a combination of a microchip implanted behind the retina of a mouse[21] and goggles with LED readouts linked to a small camera that allowed mice to distinguish black from white. This device was subsequently used by a woman who had lost her sight in a car crash, which allowed her to see object outlines and differences in light intensity. But we are some way off a device like Geordi's.

Injections in *Star Trek* are given by 'hypospray', which doesn't involve a needle and can be used through clothing. The FDA recently approved a device that can use ultrasonic waves to open pores in the skin, allowing liquids, including vaccines, to be injected without needles.[22] A device with a high-pressure jet is also being developed, which is being tested as a way to deliver vaccines in powdered form. This will mean no injections for vaccines and no need to keep vaccines at a low temperature to preserve them, which is an issue in the developing world.

Star Trek: Voyager also has an emergency hologram doctor who is an expert in all fields of medicine. Then there's the famous medical tricorder, which can diagnose diseases and collect other information about a patient by holding the device over someone's body. We are a long way off robotic doctors, although there are robots that can perform some types of surgery. In diagnosis artificial intelligence is being increasingly used, most recently in the diagnosis of breast cancer, where the technology outperformed humans.[23] Ultimately, it may well be a computer that will diagnose illness and provide treatment. There are, of course, scanners like the MRI imager, which can see inside our heads using magnetic resonance, but these are a long way from being hand-held. But there is the Standoff Patient Triage Tool (SPTT), which is being developed by the US Department of Homeland Security.[24] This device can take vital signs from up to 40 feet away – another highly clinically relevant device in the age of COVID-19. The National Space Biomedical Research Institute is developing a device that can use light to measure blood and tissue chemistry.[25] It needs to be placed on the skin, so isn't quite like the tricorder, but it's moving in that direction. There is also a device called DxtER, which can diagnose 34 conditions including diabetes, atrial fibrillation, urinary tract infection and pneumonia. It uses artificial intelligence, patient questionnaires and sensors to provide a quick assessment of a patient's health.[26] It won the top prize ($2.6m) in the annual competition for companies coming up with a device that will do what the tricorder does at a recent American Association for Clinical Chemistry meeting.

We can anticipate a large amount of progress in what medicine will look like in the future. Vast amounts of money and resources are currently being spent on medical research and the effort to develop new medicines (see Chapter 3). There is no doubt that we will continue to see tremendous progress. A useful website to keep up with all this activity is www.clinical-trials.gov, run by the National Institutes of Health in the US.[27] It currently lists 326,147 separate studies, running in all 50 states in the US and in 209 other countries. Think about that number for a minute: over 300,000 clinical trials, most of which are aiming to combat disease. In 2019 the page had 215 million views per month, with 145,000 unique visitors daily. In June 2020, over 300 trials for COVID-19 treatments were taking place. That gives you an idea of the scale of the enterprise.

The breakdown of numbers is interesting. Of the studies, 144,342 involve a new medicine being tested against a specific disease. Some of these are small molecules (tablets that are taken) and some are biologic agents that are injected: 82,880 trials are called 'behavioural' where the trial involves modifying some behaviour of the subjects on the trial (this could be dietary or life-style-based) and looking for an effect on a specific disease or condition. A further 27,041 trials are for new surgical procedures, and 32,929 involve a new medical device, usually being implanted into the bodies of the subjects involved. The number of trials running has increased four-fold since 2010.[28]

Treatment for many major diseases is often inadequate or entirely absent. These are diseases for which there is an 'unmet medical need'. They include-neurological disorders like Alzheimer's disease and Parkinson's disease, many types of cancer, heart disease, mental health disorders like depression, schizo-phrenia and psychosis, inflammatory diseases like osteoarthritis and inflamma-tory bowel disease and infectious diseases like malaria and TB and, of course, COVID-19. There are also a whole host of rare diseases (sometimes called orphan diseases because they have been neglected). Many approaches are being tested, including more recent technologies such as gene therapy, which involves replacing a faulty gene that is causing a specific disease. There are currently 25 promising trials running in gene therapy, and some therapies have already been approved. Diseases where gene therapy appears to be working include conditions that cause blindness and muscle-wasting in children.

A key question (as discussed in Chapter 3) is who will pay for these treatments, as they are extremely expensive. Given the level of activity, we are bound to see progress in most of the major diseases that continue to afflict humanity. My own research is currently trying to unravel the role metabo-lism plays in inflammatory diseases, as we've found that during inflammation, immune cells in your body burn nutrients in an unusual way. We think that targeting that deviant process might be a whole new way to treat diseases like arthritis, inflammatory bowel disease and maybe even Alzheimer's disease. I predict that we will look back on the main diseases that kill us or make us very sick currently (cancer and heart disease) as diseases of the past, with many of us living into a healthy old age. We will all die one day, and may eventually fade gracefully away, having lived a long and hopefully prosperous and fulfilled life.

What about other technologies in *Star Trek*, like the holodeck, which was invented by the time of *Star Trek: The Next Generation*? This lets people enter into a simulated virtual reality situation. Imagine if that came true? We could all end up living virtual lives. This technology is getting closer, driven mainly by the gaming industry, but businesses are also interested. Wall-to-wall high-depth monitors, sophisticated projectors, motion sensors and other technologies are becoming more commonplace. Microsoft recently announced the Illumiroom,[29] which is said to augment the space around a TV such that the boundary between the virtual and the real is blurred. There are virtual or augmented reality headsets (such as Microsoft's Hololens or Facebook's Oculus Rift), but the design isn't lightweight or comfortable for prolonged use. The holodeck is still some way off.

A whole range of other technologies in *Star Trek* are, however, close to reality. The technology that was always mentioned as coming true first was sliding doors, which in *Star Trek* were operated with ropes. These are now commonplace. Star Trek also had hand-held touchscreen 'data slates' known as personal access display devices (PADDS), which we now use in the form of iPads. The universal translator was able to scan brain waves to interpret unknown languages into the user's own language. Skype Translator Preview currently translates ten voice languages. . There are many apps that allow us to have conversations with others, using real-time translators, notably iTranslate, which has 42 languages. Google sells Pixel Buds, which are somewhat akin to the famous Babel Fish in Douglas Adams's *The Hitchhiker's Guide to the Galaxy*. In the book, these fish are placed in the ear and will translate any language.

Perhaps it will be in robotics that we will see the biggest advances in the coming decade. 'Data' is the name of the robot in *Star Trek: The Next Generation*. He is a synthetic life form with artificial intelligence, whose positronic brain allows him to be self-aware. In his early years he had trouble understanding various aspects of human behaviour and was unable to feel emotion. His creator, Dr Noonien Soong, added an 'emotion chip' to remedy this. In one episode, called 'In Theory', Data falls in love with Jenna D'Sora, a crew member, by creating a romantic subroutine for the relationship. He downloads masses of romantic novels and movies to learn about love. At one point he purposely provokes an argument with Jenna. When Jenna asks him why, he says based on his analysis of thousands of relationships, it was the optimum time for them to have a 'lovers' tiff'. Needless to say, the relationship doesn't last.

Then there's the question of sex robots.[30] Several are under development. One of the most advanced is called Harmony and is described as a 'home pleasure unit'. The industry is predicting a market size of €25 million, based on the demand for existing 'smart sex toys'. The robots are predicted to become more lifelike and will have artificial intelligence. At last, you can have an intelligent conversation with your partner before or after sex, even if he or she is a robot! There may even be different settings – compliant, argumentative, silent, Irish. A strange world is emerging when it comes to sex robots, which is raising ethical concerns. Will they make us more selfish? Will they lead to more sex addiction? Should there be limits on what they might look like (ethicists are wondering if you should be allowed to have a robot with a face like your ex)? Will they dehumanise us and lead to more deviant behaviour with real people?

We are a long way off a 'Data' although robotics is currently a much-researched area. In 2018, $4.9 billion in venture capital investment was made in 400 separate deals in the US.[31] A similar amount has been invested in China. But an Irish robot garnered a lot of attention in 2019. Robotics engineers Conor McGinn, Eamonn Bourke, Andrew Murtagh, Michael Cullinan, Cian Donovan and Niamh Donnelly from Trinity College Dublin unveiled a robot called Stevie, who has been described as Ireland's first socially assistive robot with advanced artificial intelligence.[32] Stevie was designed to support caregivers and older adults in senior living communities and other types long-term care settings. He is mobile, dextrous and can use sensing technologies like rangefinders, depth cameras and tactile and vision sensors to perceive and interact intelligently with his environment. He has also been given what the engineers who designed him call 'enhanced expressive capabilities', which facilitates more natural and intuitive communication with users.

The team consulted with a range of stakeholders during Stevie's development, including older adults, nurses and caregivers. ALONE, a major charity that supports older people in Ireland, was a key partner in its development. Stevie is now being tested in senior living communities in the US and UK. His initial jobs involved assisting care staff in performing group activities like bingo and quizzes, freeing them up to spend more and better quality-time with older adult residents. He has also been helpful in connecting

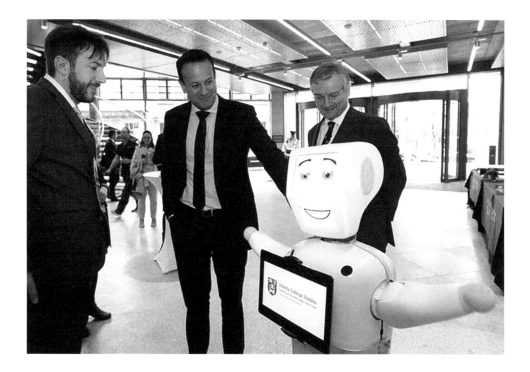

residents to their families through a video call interface that overcomes some of the usability issues with conventional tablet and smart phone applications. The experience of interacting with Stevie, enabled through an artificial intelligence system that uses advanced computer vision and language understanding algorithms, has proven to be

especially engaging, with many residents and staff taking pleasure in the conversations they have with the robot. His inventors say that a key goal for future development is to enable Stevie to engage in small talk or the craic, which will presumably make him truly Irish.

Moving away from *Star Trek* and science fiction, futurologists spend a lot of time crystal-ball gazing (not literally of course). There are a number of predictions being made to add to the ones discussed above. In the 1920s a series of books entitled *To-Day and To-Morrow* predicted the future and some of the predictions made were remarkably accurate. The mobile phone was predicted in 1924 by Archibald Low who wrote: 'In a few years' time we shall be able to chat to our friends in an aeroplane and in the streets with the help of a pocket wireless set.' J.D. Bernal, who was a Tipperary-born pioneer in the use of X-ray crystallography in molecular biology in the 1930s, predicted the World Wide Web. He wrote a book, which discussed the future, called *The World, the Flesh and the Devil*.[33] One wonders if his publisher came up with that title to ensure healthy sales. No less a figure than the eminent science fiction writer Arthur C. Clarke, writer of perhaps the greatest work of science fiction, *2001: A Space Odyssey*, called the book 'the most brilliant attempt at scientific prediction ever made'. Bernal also speculated on how just before death, the brain might be saved and transferred into a machine host. He was a fan of what became known as transhumanism – the idea that humanity should improve its species. He predicted a small sense organ for detecting wireless frequencies, enhanced vision (to allow us to perceive infra-red and ultraviolet light, and even X-rays) and ears that can detect a broader range of frequencies. He predicted how humans would be able to communicate with each other across vast differences using wireless technology. He didn't predict the computer as the main technological development of the twentieth century, which was partly because, at that time, computers (and they weren't even called that then) were run on punch cards and were analogue rather than digital. Nobody predicted the advent of electronic computers.

This, of course, gives us a warning as to how we might try and predict the future. If no-one predicted computers, which have had such a major effect on our lives, what else are we missing? And what about COVID-19? Although scientists had predicted another pandemic, nobody anticipated that. We're only now coming to terms with what COVID-19 might mean for the future. Who knows what blind spots we might have? Notwithstanding that, a timeline of what might happen over the remainder of this century has been constructed.[34] It is a likely scenario for what will happen, based on the current situation we find ourselves in. The starting point is how the world is now. Digital technology is clearly a dominant aspect of our lives with 5 billion of us (out of a total population on earth of 7 billion) having smartphones that we use constantly.[35] Twitter currently has 330 million active users, of which 66 per cent are male and 34 per cent female, and 500 million tweets are sent every day.[36] The face-with-tears-of-joy emoji alone has been used 2 billion times.[37] Facebook has 1.5 billion daily users.[38] A current key concern is the stress and anti-sociality that the overuse of smartphones is causing. There are also concerns about excessive surveillance and privacy intrusion.

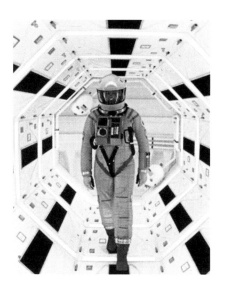

A SCENE FROM THE GREATEST SCIENCE FICTION MOVIE EVER MADE: *2001: A SPACE ODYSSEY* (1968), DIRECTED BY STANLEY KUBRICK.

What else have we got to worry about when it comes to the future? In the 2020s climate change will likely become more and more of a concern, given current trends, and will begin to threaten food and water supplies (see Chapter 13). The 2030s will finally see a substantial shift towards renewable energy supplies because of breakthroughs in nanotechnology, which will make such supplies cheaper and more efficient. Nuclear fusion will also be used more and more as a source of energy. In the 2040s the interface between genetics, nanotechnology and robotics will lead to more and more examples of transhumanism. Devices will be implanted into the human body to help combat disease, enhance our senses, allow for different forms of communication and provide entertainment. There will be colonies on Mars, and on the moon. Artificial intelligence will play a much bigger role in how businesses and governments make decisions and will supersede human decision-making. By 2060 the world's population will have reached a plateau and will start to decline. By the 2070s a full-scale environmental catastrophe will have happened, because in spite of our best efforts, climate change has still been occurring. Large-scale evacuations of cities will have happened because of rising sea levels. By the 2080s, though, scientific discovery will massively accelerate because of artificial intelligence. By the 2090s, *Homo sapiens* is no longer the dominant species on earth. The day-to-day running of countries will be done by ultra-fast, supremely intelligent robots and virtual entities. Most of the world's languages will no longer be in widespread use. English, Mandarin and Spanish will be the three dominant languages. The average employee will be working for fewer than 20 hours per week. Western Antarctica will become one of the fastest-developing regions in the world. It will have a climate similar to Alaska today, the icecaps there will have melted and immigration from regions damaged by climate change will be encouraged and incentivised. The cities there will become artistic melting pots, given the diversity of the population. How do you think that future sounds? Our children and grandchildren may well live to see it.

We humans are a curious species. We evolved from more 'primitive' life forms according to the laws of natural selection. Life is a biochemical machine. Intensely complex chemistry, the right conditions, the dinosaurs being wiped out by an asteroid (which made room for our ancestor, a small shrew-like creature, to thrive) and a long, long time was all that was needed to get to us. Our curiosity led to us inventing science and we found out a large number of interesting things about the world we live in. We continue to make discoveries at an incredible rate, enhanced by the machines we invented, especially in the past ten years or so as the digital age has taken off. As the cosmologist and science communicator Carl Sagan once said: 'Somewhere, something incredible is waiting to be known.' We are the only species (as far as we can tell) who can look inward and figure out what makes us tick. And then we use all this science to invent new technologies, few of which most of us understand. A curious business indeed, that will continue apace as we head towards an uncertain future.

When I was training to be a research scientist in Cambridge in the late eighties I was working for an outstanding scientist and rheumatologist called

Jerry Saklatvala, who is my most important mentor. He was working on a protein made by your immune system called TNF. At that time, TNF was linked to cancer. As often happens in science, as more experiments are done, this

link turned out to be limited. But Jerry found that when TNF was added to cartilage prepared from pig's trotters (a good source of cartilage) it caused the cartilage to be destroyed. We thought, hmmm ... Jerry's discovery might be interesting. Cartilage breakdown is a key feature of rheumatoid arthritis and was the reason why people ended up with bent fingers, using walking sticks or in wheelchairs. It is a disease which left untreated unrelentingly eats away at your joints. Medical research has many false dawns. Although Jerry's findings were interesting, few dared hope that TNF would turn out to be an important target for medicines to treat rheumatoid arthritis. But that is exactly what has happened. The medicines that block TNF stop joint destruction and help many millions of people, preventing it from being a painful disease. Jerry's observation turned out to be very important for the development of these new medicines. Who knew? And who knows where all the research that's going on right now in labs all over the world will lead?

My final bottom line: I remain optimistic that things will continue to improve for humanity. Let's keep that dream alive.

I can't wait to see what happens next.

ENDNOTES

INTRODUCTION

1   World Health Organization (2019). Tobacco explained: the truth about the tobacco industry ... in its own words. Available at: https://www.who.int/tobacco/media/en/TobaccoExplained.pdf

2   R. Matthews (2000) *Storks deliver babies* (p= 0.008). Teaching Statistics, 22(2): 36–38.

3   Snopes (2016). Does this map show mad cow disease prevalence vs. Brexit voters? Available at: https://www.snopes.com/fact-check/mad-cow-versus-brexit

CHAPTER 1

1   M. McKenna and D. Pereboo (2016). *Free Will* (Routledge Contemporary Introductions to Philosophy). New York: Routledge.

2   R. Pippin (2012). *Introductions to Nietzsche*. Cambridge University Press.

3   A. Vilenkin and M. Tegmark (2011). The case for parallel universes. *Scientific American*, 19 July.

4   B. Gholipour (2019). Philosophers and neuroscientists join forces to see whether science can solve the mystery of free will. *Science*, 21 March. Available at: https://www.sciencemag.org/news/2019/03/philosophers-and-neuroscientists-join-forces-see-whether-science-can-solve-mystery-free

5   B. Libet et al. (1983). Time of conscious intention to act in relation to onset of cerebral activity (readiness-potential). The unconscious initiation of a freely voluntary act. *Brain*, 106: 623–642.

6   W.R. Klemm (2010). Free will debates: simple experiments are not so simple. *Advances in Cognitive Psychology*, 6: 47–65.

7   R.H. Anderberg (2016). The stomach-derived hormone ghrelin increases impulsive behavior. *Neuropsychopharmacology*, 41: 1199–1209.

8   J. Skrynka and B.T. Vincent (2019). Hunger increases delay discounting of food and non-food rewards. *Psychonomic Bulletin and Review*, 26: 1729–1737.

9   M. Reynolds (2014). When you should never make a decision. *Psychology Today*, 17 April.

10   A. Vyas et al. (2007). Behavioral changes induced by *Toxoplasma* infection of rodents are highly specific to aversion of cat odors. *Proceedings of the National Academy of the Sciences of the USA*, 104: 6442–6644.

11   J. Flegr (2007). Effects of toxoplasma on human behavior. *Schizophrenia Bulletin*, 33: 757–760.

12   A. Stock (2017). Humans with latent toxoplasmosis display altered reward modulation of cognitive control. *Scientific Reports*, 7: 10170.

13   C. Dixon (2018). How much control do we really have over how we think and act? *Irish Examiner*, 11 January.

14   J. Lindova et al. (2006). Gender differences in behavioural changes induced by latent toxoplasmosis. *International Journal for Parasitology*, 36: 1485–1492.

15   Better Explained (n.d.). Understanding the birthday paradox. Available at: https://betterexplained.com/articles/understanding-the-birthday-paradox/

16   V. Jessop (2007). *Titanic Survivor: The Memoirs of Violet Jessop, Stewardess*. History Press.

17   G. Adams (2006). How to live your life by numbers. *The Independent*, 26 November.

18   R. Gillett and I. De Luce (2019). Science says parents of successful kids have these 23 things in common. *Business Insider*, 23 May. Available at: https://www.businessinsider.com/how-parents-set-their-kids-up-for-success-2016-4?r=US&IR=T

19   Harvard Business School (2015). Having a working mother is good for you. 18 May. Available at: https://www.hbs.edu/news/releases/Pages/having-working-mother.aspx

20   E.J. Dixon-Roman et al. (2013). Race, poverty and SAT scores: modeling the influences of family income on black and white high school students' SAT performance. *Teachers College Record*, 115(4).

21   T.R. Mitchell et al. (2003). 'Motivation' in Walter C. Borman et al. (eds), *Handbook of Psychology*, Vol. 12. John Wiley & Sons.

22   T.N. Robinson et al. (2007). Effects of fast-food branding on young children's taste preferences. *Archives of Pediatric and Adolescent Medicine*, 161: 792–797.

23   L. Donnelly (2019). Junk food giants must stop marketing to children – or see their ads banned entirely, says health chief. *The Telegraph*, 14 March.

24   D. Campbell (2017). Children seeing up to 12 adverts for junk food an hour on TV, study finds. *The Guardian*, 28 November.

25   World Health Organization (2019). Reducing the impact of marketing of foods and non-alcoholic beverages on children. Available at: https://www.who.int/elena/titles/food_marketing_children/en/

26   S. Boseley (2016). Junk food ads targeting children banned in non-broadcast media. *The Guardian*, 8 December.

27   J. Shannon (2018). Majority favour ban on junk food advertising to kids. *Irish Heart Foundation Newsletter*, 7 November.

28   E. Ring (2018). Junk food adverts have become a monster. *Irish Examiner*, 8 November.

29   C.C. Steele et al. (2017). Diet-induced impulsivity: effects of a high-fat and a high-sugar diet on impulsive choice in rats. *PLOS One* 12, e0180510.

30   D. Lynkova (2019). Key smartphone addiction statistics. *Leftronic*. Available at: https://leftronic.com/smartphone-addiction-statistics/

31   S.C. Matz (2017). Psychological targeting as an effective approach to digital mass persuasion. *Proceedings of the National Academy of the Sciences of the USA*, 114: 12714–12719.

32   R. Verkaik (2018). Cambridge Analytica: inside the murky world of swinging elections and advising

dictators. *iNews*, 23 March (updated 6 September 2019). Available at: https://inews.co.uk/news/technology/cambridge-analytica-facebook-data-protection-312276.

33   J. Doward *and A. Gibbs* (2017). Did Cambridge Analytica influence the Brexit vote and the US election? *The Guardian*, 4 March.

34   N. Lomas (2018). Facebook finally hands over Leave campaign Brexit ads. *Techcrunch*, 26 July.

35   J. Doward and A. Gibbs.

CHAPTER 2

1   K. Mills and D. Ahlstrom (2019). Vaccines: a life-saving choice. *Royal Irish Academy Expert Statement, Life and Medical Sciences Committee.*

2   World Health Organization (2019). Immunization. 5 December. Available at: https://www.who.int/news-room/facts-in-pictures/detail/immunization

3   J.L. Goodson. and J.F. Seward (2015). Measles 50 years after use of measles vaccine. *Infectious Disease Clinics of North America*, 29(4):725–743.

4   World Health Organization (2019) Ten threats to global health in 2019. Available at: https://www.who.int/emergencies/ten-threats-to-global-health-in-2019

5   M. Ferren et al. (2019). Measles encephalitis: towards new therapeutics. *Viruses*, 11(11).

6   H. Wang et al. (2016). Global, regional, and national levels of maternal mortality, 1990–2015: a systematic analysis for the global burden of disease study 2015. *Lancet.* 388(10053): 1775–1812.

7   N. Nathanson and O.M. Kew (2010). From emergence to eradication: the epidemiology of poliomyelitis deconstructed. *American Journal of Epidemiology*, 172(11):1213–1229.

8   Centers for Disease Control and Prevention (2019). Polio elimination in the United States. Available at: https://www.cdc.gov/polio/what-is-polio/polio-us.html

9   I. Grundy (2019). Montagu, Lady Mary Wortley. *Oxford Dictionary of National Biography.* Oxford University Press.

10   J.R. Smith (2006). Jesty, Benjamin. *Oxford Dictionary of National Biography.* Oxford University Press.

11   G. Williams (2019). The original anti-vaxxers. *The Economist 1843*, 30 August.

12   M. Arbyn et al. (2018). Prophylactic vaccination against human papillomaviruses to prevent cervical cancer and its precursors. *Cochrane Database of Systematic Reviews*, 5: CD009069.

13   J. Zheng et al. (2019). Prospects for malaria vaccines: pre-erythrocytic stages, blood stages, and transmission-blocking stages. *BioMed Research International*, 2019:9751471.

14   D. Malvy et al. (2019). Ebola virus disease. *Lancet*, 18: 936–948.

15   F. Amanat and F. Krammer (2020). SARS-CoV-2 vaccines: status report. *Immunity.* pii: S1074-7613(20)30120. Available at: https://www.ncbi.nlm.nih.gov/pubmed/32259480

16   A.J. Young (2019). Adjuvants: what a difference 15 years makes! *Veterinary Clinics of North America: Food Animal Practice*, 35(3): 391-403. doi: 10.1016/j.cvfa.2019.08.005.

17   S. Marsh (2018). Take-up of MMR vaccine falls for fourth year in a row in England. *The Guardian*, 18 September.

18   National Vaccine Injury Compensation Program. *Health Resources and Services Administration.* Available at: https://www.hrsa.gov/vaccine-compensation/index.html

19   Centers for Disease Control and Prevention (2014). Report shows 20-year US immunization program spares millions of children from disease. Available at: https://www.cdc.gov/media/releases/2014/p0424-immunization-program.html

20   Centers for Disease Control and Prevention (2019). Q&As about vaccination options for preventing measles, mumps, rubella, and varicella. Available at: https://www.cdc.gov/vaccines/vpd/mmr/hcp/vacopt-faqs-hcp.html

21   Centers for Disease Control and Prevention (n.d.). Measles, mumps, and rubella diseases and how to protect against them. Available at: https://www.cdc.gov/vaccinesafety/vaccines/mmr-vaccine.html

22   N.P. Klein et al. (2012). Safety of quadrivalent human papillomavirus vaccine administered routinely to females. *Archives of Pediatric and Adolescent Medicine* 166(12):1140–1148. doi: 10.1001/archpediatrics.2012.1451

23   D. Gorski (2010). The fall of Andrew Wakefield. *Science-Based Medicine*, 22 February. Available at: https://sciencebasedmedicine.org/the-fall-of-andrew-wakefield/

24   American Academy of Pediatrics (n.d.). American Academy of Pediatrics urges parents to vaccinate children to protect against measles. Available at: https://www.aap.org/en-us/about-the-aap/aap-press-room/Pages/American-Academy-of-Pediatrics-Urges-Parents-to-Vaccinate-Children-to-Protect-Against-Measles.aspx

25   L.E. Taylor et al. (2016). Vaccines are not associated with autism: an evidence-based meta-analysis of case-control and cohort studies. *Vaccine* 34: 3223–3224.

26   T. Leonard (2019) Rewards for the High Priest of MMR hysteria. *Daily Mail*, 10 October. Available at: https://www.dailymail.co.uk/news/article-7556279/Andrew-Wakefield-struck-anti-MMR-science-millionaire-lifestyle.html

27   S. Pollak (2019). Number of Irish measles cases more than triples between 2017 and 2018. *Irish Times*, 25 April. Available at: https://www.

irishtimes.com/news/health/number-of-irish-measles-cases-more-than-triples-between-2017-and-2018-1.3871238

28 H. Holzmann (2016). Eradication of measles: remaining challenges. *Medical Microbioogy and Immunology*, 205(3):201–208. doi: 10.1007/s00430-016-0451-4.

29 F. Rahimi and Amin Talebi Bezmin Abadi (2020). Practical strategies against the novel coronavirus and COVID-19 – the imminent global threat. *Archives of Medical Research*, S0188-4409(2) 30287–3. Available at: https://www.sciencedirect.com/science/article/abs/pii/S0188440920302873#!

30 Z. Horne et al. (2015). Countering anti-vaccination attitudes. *Proceedings of the National Academy of Science of the USA* 112: 10321–10324.

31 C.A. Bonville et al. (2017). Immunization attitudes and practices among family medicine providers. *Human Vaccines and Immunotherapeutics*, 13: 2646–2653.

32 M.F. Daley and J.M. Glanz (2011). Straight talk about vaccination. *Scientific American*, 1 September.

CHAPTER 3

1 D. Stipp (2013). Is fasting good for you? *Scientific American*, 308(1): 23–24.

2 W.F. Pirl and A.J. Roth (1999). Diagnosis and treatment of depression in cancer patients. *Cancer Network*, 13(9): 1293–1301.

3 S. Rezaei et al. (2019). Global prevalence of depression in HIV/AIDS: a systematic review and meta-analysis. *BMJ Supportive and Palliative Care*, 9: 404–412.

4 T. Sullivan (2019). A tough road: cost to develop one new drug is $2.6 billion; approval rate for drugs entering clinical development is less than 12%. *Policy and Medicine*, 21 March.

5 A. Nieto-Rodriguez (2017). Is the iPhone the best project in history? *CIO*. Available at: https://www.cio.com/article/3236171/is-the-iphone-the-best-project-in-history.html

6 S. Held et al. (2009). Impact of big pharma organizational structure on R&D productivity. *Schriften zur Gesundheitsoekonmie* 17.

7 UK Medical Research Council (n.d.). Facts & figures. Available at: https://mrc.ukri.org/about/what-we-do/spending-accountability/facts/

8 E.J. Emanuel (2019). Big pharma's go-to defense of soaring drug prices doesn't add up. Just how expensive do prescription drugs need to be to fund innovative research? *The Atlantic*. Available at: https://www.theatlantic.com/health/archive/2019/03/drug-prices-high-cost-research-and-development/585253/

9 Blass, B. (2015). *Basic Principles of Drug Discovery and Development*. Elsevier.

10 Hilt, P.J. (2003). *Protecting America's Health: The FDA, Business, and One Hundred Years of Regulation*. Random House.

11 DTS Language Services (2018). How much does it cost to run a clinical trial? Available at: https://www.dtstranslates.com/clinical-trials-translation/clinical-trial-cost/

12 I. Torjesen (2015). Drug development: the journey of a medicine from lab to shelf. *Pharmaceutical Journal*. Available at: https://www.pharmaceutical-journal.com/publications/tomorrows-pharmacist/drug-development-the-journey-of-a-medicine-from-lab-to-shelf/20068196.article?firstPass=false

13 Biotechnology Innovation Organization (n.d.). Clinical Development Success Rates 2006–2015. Available at: https://www.bio.org/sites/default/files/legacy/bioorg/docs/Clinical%20Development%20Success%20Rates%202006-2015%20-%20BIO,%20Biomedtracker,%20Amplion%202016.pdf

14 R. Imai Takebe, S. Ono et al. (2018). The current status of drug discovery and development as originated in United States academia: The influence of industrial and academic collaboration on drug discovery and development. *Clinical and Translational Science*, 11(6).

15 C. Hale (2018) New MIT study puts clinical research success rate at 14 percent. *CenterWatch*. Available at: https://www.centerwatch.com/articles/12702-new-mit-study-puts-clinical-research-success-rate-at-14-percent

16 R. Bazell (1998). *Her-2: The Making of Herceptin, a Revolutionary Treatment for Breast Cancer*. Random House.

17 P.D. Risse (2017). Bet on biomarkers for better outcomes. *Life Science Leader*, 6 April.

18 Leber congenital amaurosis. *Genetics Home Reference*.

19 A.M. Maguire et al. (2008). Safety and efficacy of gene transfer for Leber's congenital amaurosis. *New England Journal of Medicine*, 358: 2240–2248.

20 Institute for Clinical and Economic Review (2018). Final Report: Broader benefits of Voretigene Neparvovec to affected individuals and society provide reasonable long-term value despite high price. Available at: https://icer-review.org/announcements/voretigene-final-report/

21 M.E Condren and M. Bradshaw (2013). Ivacaftor: a novel gene-based therapeutic approach for cystic fibrosis. *Journal of Pediatric Pharmacology and Therapeutics*, 18: 8–13.

22 L.B. Feng et al. (2018). Precision medicine in action: the impact of Ivacaftor on cystic fibrosis-related hospitalizations. *Health Affairs (Millwood)*, 37: 773-779.

23 D. Cohen and J. Raftery (2014). Paying twice: the 'charitable' drug with a high price tag. *British Medical Journal*, 348: 18–21.

24 J. Fauber (2013). Cystic fibrosis: charity and industry partner for profit. *MedPage Today*. Available at: https://www.medpagetoday.com/pulmonology/cysticfibrosis/39217

25 B. Fidler (2014). CF Foundation cashes out

on Kalydeco in $3.3B sale to Royalty Pharma. *Xconomy*. Available at: https://xconomy.com/boston/2014/11/19/cf-foundation-cashes-out-on-kalydeco-in-3-3b-sale-to-royalty-pharma/

26   D. Sharma et al. (2018). Cost-effectiveness analysis of Lumacaftor and Ivacaftor combination for the treatment of patients with cystic fibrosis in the United States. *Orphanet Journal of Rare Diseases*, 13: 172.

27   Orkambi monograph. *Drugs*. Available at: https://www.drugs.com/monograph/orkambi.html

28   T. Ferkol and P. Quinton. (2015). Precision medicine: at what price? *American Journal of Respiratory and Critical Care Medicine*. 196, 15 September.

29   S.M. Hoy. (2019). Elexacaftor/Ivacaftor/Tezacaftor: first approval. *Drugs*, 79: 2001–2007.

30   Advisory Board (2019). FDA approves drug to treat cystic fibrosis in patients 12 and older – and it will cost $311,503. Available at: https://www.advisory.com/daily-briefing/2019/10/28/cf-drug

31   P. Cullen (2019). HSE agrees to reimburse cost of new cystic fibrosis treatment. *Irish Times*, 13 December.

32   I. Shahid (ed.) (2018). *Hepatitis C: From Infection to Cure*. InTechOpen.

33   M. Goozner (2014). Why Sovaldi shouldn't cost $84,000. *Modern Healthcare*, 44(18): 26.

34   E. Hafez (2018). A new potent NS5A inhibitor in the management of hepatitis C virus: Ravidasvir. *Current Drug Discovery Technologies*, 15(1): 24–31.

35   World Health Organization (2017). Close to 3 million people access hepatitis C cure. Available at: https://www.who.int/news-room/detail/31-10-2017-close-to-3-million-people-access-hepatitis-c-cure

36   M. Costanzo et al. (2020) SARS-CoV-2: recent reports on antiviral therapies based on Lopinavir/Ritonavir, Darunavir/Umifenovir, Hydroxychloroquine, Remdesivir, Favipiravir and other drugs for the treatment of the new coronavirus. *Current Medicinal Chemistry*. doi: 10.2174/09298673276662004161311117.

37   R.G. Frank and L.M. Nichols (2019). Medicare drug-price negotiation – why now … and how. *New England Journal of Medicine* 381: 1404–1406.

38   Drugs.com (2018). EpiPen costs and alternatives: what are your best options? Available at: https://www.drugs.com/article/epipen-cost-alternatives.html

39   S. Gordon (2016). Cost of insulin rises threefold in just a decade: study. *HealthDay*. Available at: https://consumer.healthday.com/diabetes-information-10/insulin-news-414/cost-of-insulin-rises-threefold-in-just-a-decade-study-709697.html

40   L. Entis (2019). Why does medicine cost so much? Here's how drug prices are set. *Time*, 9 April.

41   P. Cullen (2019). Irish patients pay 'six times global average' for generic drugs. Cost of branded drugs almost 14% below average in 50 countries surveyed. *Irish Times*, 21 November.

42   E. Edwards (2018). Ireland urgently needs access to new drugs, says pharma body. *Irish Times*, 22 June.

43   E. Ring (2018). Warning to control costs of medicines. *Irish Examiner*, 7 April.

44   US Food and Drug Administration (2018). 2018 New Drug Therapy Approvals. Available at: https://www.fda.gov/files/drugs/published/New-Drug-Therapy-Approvals-2018_3.pdf

45   R. Stein (2018). At $2.1 million, new gene therapy is the most expensive drug ever. *NPR*, 24 May 2019. Available at: https://www.npr.org/sections/health-shots/2019/05/24/725404168/at-2-125-million-new-gene-therapy-is-the-most-expensive-drug-ever

46   World Health Organization (2019). WHO Model List of Essential Medicines, 21st List. Available at: https://www.who.int/medicines/publications/essentialmedicines/en/

CHAPTER 4

1   J. Clarke (2018). Weight on the mind … *Irish Health*. Available at: http://www.irishhealth.com/article.html?id=2354

2   T. O'Brien (2019). Two-thirds of men in Ireland are overweight or obese, report finds. *Irish Times*, 20 November.

3   Health Service Executive (n.d.) Healthy Eating and Active Living Programme. Available at: https://www.hse.ie/eng/about/who/healthwellbeing/our-priority-programmes/heal/

4   A. Harris (2018). Obesity in Irish men increasing at 'alarming' rate. *Irish Times*. 5 September.

5   World Health Organization (2020). Obesity and overweight. Available at: https://www.who.int/news-room/fact-sheets/detail/obesity-and-overweight

6   K. Donnelly (2015). *Adolphe Quetelet, Social Physics and the Average Men of Science, 1796–1874*. University of Pittsburgh Press.

7   N. Rasmussen (2019). Downsizing obesity: on Ancel Keys, the origins of BMI, and the neglect of excess weight as a health hazard in the United States from the 1950s to 1970s. *Journal of the History of the Behavioral Sciences*, 55: 299–318.

8   X. Pi-Sunyer (2009). The medical risks of obesity. *Postgraduate Medical Journal* 121: 21–33.

9   PSC Secretariat (2009) Body-mass index and cause-specific mortality in 900,000 adults: collaborative analyses of 57 prospective studies. *Lancet* 373: 1083–1096.

10   L. Donnelly (2019). Obesity overtakes smoking as the leading cause of four major cancers. *Daily Telegraph*, 3 July.

11   N. Devon (2017). You are your looks: that's what society tells girls. No wonder they're depressed. *The Guardian*, 22 September.

12   K. Miller (2015). Sad proof that most women don't think they're beautiful. *Women's Health*, 7 April.

13 DoSomething.org (n.d.). 11 facts about body image. Available at: https://www.dosomething.org/us/facts/11-facts-about-body-image

14 K. Pallarito (2016). Many men have body image issues, too. *WebMD*. Available at: https://www.webmd.com/men/news/20160318/many-men-have-body-image-issues-too#1

15 C. Markey (2019). Teens, body image, and social media. *Psychology Today*, 14 February.

16 F. Rubino et al. (2020). Joint international consensus statement for ending stigma of obesity. *Nature Medicine* 26: 485–497. doi: 10.1038/s41591-020-0803-x. Available at: https://www.nature.com/articles/s41591-020-0803-x.

17 WebMD (n.d.). Estimated calorie requirements. Available at: https://www.webmd.com/diet/features/estimated-calorie-requirements

18 J.M. Friedman (2019). Obesity is in the genes. *Scientific American Blog*, 31 October. Available at: https://blogs.scientificamerican.com/observations/obesity-is-in-the-genes/

19 V.V. Thaker (2017). Genetic and epigenetic causes of obesity. *Adolescent Medicine: State of the Art Reviews*, 28(2): 379–405.

20 C.T. Montague et al. (1997). Congenital leptin deficiency is associated with severe early-onset obesity in humans. *Nature*, 387: 903–908.

21 A.E. Locke et al. (2015). Genetic studies of body mass index yield new insights for obesity biology. *Nature*, 518: 197–206.

22 S. Kashyap et al. (2010). Bariatric surgery for type 2 diabetes: weighing the impact for obese patients. *Cleveland Clinic Journal of Medicine*, 77: 468–476.

23 R.B. Kumar and L.J. Aronne (2017). Pharmacologic treatment of obesity, in K.R. Feingold, B. Anawalt, A. Boyce et al. (eds), *Endotext*, South Dartmouth (MA). Available at: https://www.ncbi.nlm.nih.gov/books/NBK279038.

24 G. Cheyne (1724). *An Essay of Health and Long Life*. George Strahan.

25 W. Banting (1864). *Letter on Corpulence, Addressed to the Public*.

26 L.H. Peters (1918). *Diet and Health: With Key to the Calories*. Reilly and Lee.

27 Boston Medical Center (n.d.). Weight management. Available at: https://www.bmc.org/nutrition-and-weight-management/weight-management

28 ABC News (2005). Oprah calls her biggest moment a big mistake. 11 November. Available at: https://abcnews.go.com/GMA/story?id=1299232

29 J. Owen (2010). Human meat: just another meal for early Europeans? *National Geographic*, 2 September.

30 J.J. Hidalgo-Mora et al. (2020). The Mediterranean diet: a historical perspective on food for health. *Maturitas*, 132: 65–66.

31 E. Rillamas-Sun et al. (2014). Obesity and survival to age 85 years without major disease or disability in older women. *JAMA Internal Medicine* 174: 98–106.

32 Harvard Medical School (2017). Abdominal obesity and your health. Available at: https://www.health.harvard.edu/staying-healthy/abdominal-obesity-and-your-health

33 K.A. Scott et al. (2012). Effects of chronic social stress on obesity. *Current Obesity Reports*, 1: 16–25.

34 A. Astrup (2000). The role of low-fat diets in body weight control: a meta-analysis of ad libitum dietary intervention studies. *International Journal of Obesity and Related Metabolic Disorders*, 24: 1545–1552.

35 L.F. Donze and L.J. Cheskin (2003). Obesity treatment. *Encyclopedia of Food Sciences and Nutrition* (2nd edition), 4232–4240.

36 Mayo Clinic (2017). Low-carb diet: can it help you lose weight? Available at: https://www.mayoclinic.org/healthy-lifestyle/weight-loss/in-depth/low-carb-diet/art-20045831

37 C. Duraffourd et al. (2012). Mu-opioid receptors and dietary protein stimulate a gut-brain neural circuitry limiting food intake. *Cell*, 150: 377–388.

38 A.N. Friedman (2004). High-protein diets: potential effects on the kidney in renal health and disease. *American Journal of Kidney Diseases*, 44: 950–962.

39 C.B. Ebbeling et al. (2018). Effects of a low carbohydrate diet on energy expenditure during weight loss maintenance: randomised trial. *British Medical Journal*, 363:k4583.

40 E. Finkler et al. (2012). Rate of weight loss can be predicted by patient characteristics and intervention strategies. *Journal of the Academy of Nutrition and Diet*, 112: 75–80.

41 National Heart, Lung and Blood Institute (1998). Clinical guidelines on the identification, evaluation, and treatment of overweight and obesity in adults. The evidence report. NIH Publication No. 98–4083.

42 2-4-6-8 Diet. *Ana Diets*. http://anadiets.blogspot.com/2008/12/2-4-6-8-diet.html

43 Weight Watchers (n.d.). About us – history and philosophy. Available at: https://www.weightwatchers.com/about/his/history.aspx

44 K.A. Gudzune et al. (2015). Efficacy of commercial weight loss programs: an updated systematic review. *Annals of Internal Medicine*, 162: 501–512.

45 Z.J. Ward et al. (2019). Projected U.S. state-level prevalence of adult obesity and severe obesity. *New England Journal of Medicine*, 381: 2440–2450.

CHAPTER 5

1   J. Menasche Horowitz and N. Graf (2019). Teens see anxiety and depression as a major problem among their peers. *Pew Research Centre*, 20 February 2019. Available at: https://www.pewsocialtrends.org/2019/02/20/most-u-s-teens-see-anxiety-and-depression-as-a-major-problem-among-their-peers/

2   Anxiety and Depression Association of America (n.d.). Facts and statistics. Available at: https://adaa.org/about-adaa/press-room/facts-statistics

3   C. O'Brien (2019). Mental health: record numbers of third-level students seek help. *Irish Times*, 17 June.

4   A.K. Ibrahim et al. (2013). A systematic review of studies of depression prevalence in university students. *Journal of Psychiatric Research*, 47: 391–400.

5   M. Casey Olseth (2018). Is success a risk factor for depression? *Op-Med Doximity*. Available at: https://opmed.doximity.com/articles/is-success-a-risk-factor-for-depression.

6   J.W. Barnard, (2009). Narcissism, over-optimism, fear, anger and depression: the interior lives of corporate leaders. *University of Cincinnati Law Review*, vol. 77: 405–430.

7   WebMD (2018). Depression diagnosis. Available at: https://www.webmd.com/depression/guide/depression-diagnosis

8   I. Kirsch (2019). Placebo effect in the treatment of depression and anxiety. *Frontiers in Psychiatry*. Available at: https://doi.org/10.3389/fpsyt.2019.00407

9   O. Renick (2011). France, U.S. have highest depression rates in world, study finds. *Bloomberg*, 25 July. Available at: https://www.bloomberg.com/news/articles/2011-07-26/france-u-s-have-highest-depression-rates-in-world-study-suggests

10   C.T. Beck et al. (2006). Further development of the postpartum depression predictor inventory revised. *Journal of Obstetric, Gynecologic and Neonatal Nursing*, 35(6): 735–745.

11   M.L. Scott (1983). Ventricular enlargement in major depression. *Psychiatry Research*, 8(2): 91–3.

12   E. Bulmore (2018). *The Inflamed Mind*. Picador.

13   ClinCalc (n.d.). Fluoxetine hydrochloride drug usage statistics, United States, 2007–2017. Available at: https://clincalc.com/DrugStats/Drugs/FluoxetineHydrochloride

14   G. Iacobucci (2019). NHS prescribed record number of antidepressants last year. *British Medical Journal* 364: I15508.

15   S. McDermott (2018). HSE prescriptions for antidepressants and anxiety medications up by two thirds since 2009. *The Journal*, 1 August. Available at: https://www.thejournal.ie/ireland-antidepressant-anxiety-medicine-prescriptions-4157452-Aug2018/

16   S. Borges et al. (2014). Review of maintenance trials for major depressive disorder: a 25-year perspective from the US Food and Drug Administration. *Journal of Clinical Psychiatry*, 75(3): 205–14. doi: 10.4088/JCP.13r08722.

17   A. Cipriani et al. (2018). Comparative efficacy and acceptability of 21 antidepressant drugs for the acute treatment of adults with major depressive disorder: a systematic review and network meta-analysis. *The Lancet* 391, 1357–1366.

18   Harvard Health Publishing (2019). What causes depression? Available at: https://www.health.harvard.edu/mind-and-mood/what-causes-depression

19   E. Palmer et al. (2019). Alcohol hangover: underlying biochemical, inflammatory and neurochemical mechanisms. *Alcohol* 1; 54(3): 196–203.

20   F.W. Lohoff (2010). Overview of the genetics of major depressive disorder. *Current Psychiatry Reports*, 12(6): 539–546.

21   M.M. Weissman et al. (2005). Families at high and low risk for depression: a 3-generation study. *Archives of General Psychiatry*, 62(1): 29–36.

22   D.M. Howard et al. (2019). Genome-wide meta-analysis of depression identifies 102 independent variants and highlights the importance of the prefrontal brain regions. *Nature*, 22: 343–352.

23   N.R. Wray et al. (2018). Genome-wide association analyses identify 44 risk variants and refine the genetic architecture of major depression. *Nature Genetics*, 50(5): 668–681.

24   S.K. Adams and T.S. Kisler (2013). Sleep quality as a mediator between technology-related sleep quality, depression and anxiety. *Cyberpsychology, Behavior and Social Networking*, 16(1): 25–30. doi: 10.1089/cyber.2012.0157.

25   E. Driessen and S.D. Hollon (2010). Cognitive behavioral therapy for mood disorders: efficacy, moderators and mediators psychiatry. *Medical Clinics of North America*, 33(3): 537–555.

26   R. Haringsma et al. (2006). Effectiveness of the Coping With Depression (CWD) course for older adults provided by the community-based mental health care system in the Netherlands: a randomized controlled field trial. *International Psychogeriatrics*, 18(2): 307–325.

27   J. Spijker et al. (2002). Duration of major depressive episodes in the general population: results from the Netherlands general population: results from the Netherlands Mental Health Survey and Incidence Study (NEMESIS). *British Journal of Psychiatry* 181: 208–213.

28   US Department of Health and Human Services (n.d.). Does depression increase the risk of suicide? Available at: https://www.hhs.gov/answers/mental-health-and-substance-abuse/does-depression-increase-risk-of-suicide/index.html

29   Centers for Disease Control and Prevention (2015). Suicide – facts at a glance. Available at: https://www.cdc.gov/violenceprevention/pdf/suicide-datasheet-a.pdf

30   S. Thibault (2018). Suicide is declining almost everywhere. *The Economist*, 24 November.

31   J. Menasche Horowitz and N. Graf (2019). Teens see anxiety and depression as a major problem among their peers. *Pew Research Centre*, 20 February 2019. Available at: https://www.pewsocialtrends.org/2019/02/20/most-u-s-teens-see-anxiety-and-depression-as-a-major-problem-among-their-peers/

32   A. O'Donovan (2013). Suicidal ideation is associated with elevated inflammation in patients with major depressive disorder. *Depression and Anxiety*, 30: 307–314.

33   J.R. Kelly et al. (2016) Transferring the blues: depression associated gut microbiota induces neurobehavioral changes in the rat. *Journal of Psychiatric Research*, 82: 109–118.

34   F.S. Correia-Melo et al. (2020). Efficacy and safety of adjunctive therapy using esketamine or racemic ketamine for adult treatment-resistant depression: a randomized, double-blind, non-inferiority study. *Journal of Affective Disorders*, 264: 527–534.

35   J. Lawrence (2015). The secret life of ketamine. *Pharmaceutical Journal*, 21/28 March, 294(7854/5). doi 10.1211/PJ.2015.20068151.

36   S.B. Goldberg et al. (2020). The experimental effects of psilocybin on symptoms of anxiety and depression: a meta-analysis. *Psychiatry Research*, 284:112749

37   SR Chekroud et al. (2018). Association between physical exercise and mental health in 1·2 million individuals in the USA between 2011 and 2015: a cross-sectional study. Lancet Psychiatry. 5(9):739-74.

CHAPTER 6

1   C. Pope (2019). Typical smartphone user in Ireland checks device 50 times a day. *Irish Times*, 4 December.

2   S. Johnson (2019). Almost a third of teenagers sleep with their phones, survey finds. *Edsource*, 28 May. Available at: https://edsource.org/2019/almost-a-third-of-teenagers-sleep-with-their-phones-survey-finds/612995

3   Business2Community (2014). 89% of us have PPV syndrome and we don't even know it. Available at: https://www.business2community.com/mobile-apps/89-us-ppv-syndrome-dont-even-know-0757768

4   RescueTime Blog (2018). Here's how much you use your phone during the workday. Available at: https://blog.rescuetime.com/screen-time-stats-2018/

5   *Imaging Technology News* (2018). Smartphone addiction creates imbalance in brain. 11 January. Available at: https://www.itnonline.com/content/smartphone-addiction-creates-imbalance-brain

6   S.S. Alavi et al. (2012). Behavioral addiction versus substance addiction: correspondence of psychiatric and psychological views. *International Journal of Preventive Medicine*, 3: 290–294.

7   M.G. Griswold et al. (2018). Alcohol use and burden for 195 countries and territories, 1990–2016: a systematic analysis for the Global Burden of Disease Study 2016. *Lancet*. 392, 1015–1035.

8   World Health Organization (2014). Global status report on alcohol and health: country profiles. *World Health Organization*. Available at: https://www.who.int/substance_abuse/publications/global_alcohol_report/msb_gsr_2014_2.pdf?ua=1

9   L. Delaney et al. (2013). Why do some Irish drink so much? Family, historical and regional effects on students' alcohol consumption and subjective normative thresholds. *Review of Economics of the Household*, 11: 1–27.

10   C. Feehan (2020). More women than men reporting benzo and opiate use when seeking help for alcohol addiction. *Irish Independent*, 1 February.

11   V. Preedy (2019). *Neuroscience of Nicotine. Mechanisms and Treatment* (1st edition). Academic Press.

12   *Healthy Ireland Summary Report 2019*. Irish Government Publications. Available at: https://assets.gov.ie/41141/e5d6fea3a59a4720b081893e11fe299e.pdf

13   HRB National Drugs Library. *Health Research Board*. Available at: https://www.drugsandalcohol.ie/30619/

14   E.J. Nesteler (2005). The neurobiology of cocaine addiction. *Science and Practice Perspectives*, 3(1): 4–10.

15   G. Battaglia et al. (1990). MDMA effects in brain: pharmacologic profile and evidence of neurotoxicity from neurochemical and autoradiographic studies. In S.J. Peroutka (ed.), *Ecstasy: The Clinical, Pharmacological and Neurotoxicological Effects of the Drug MDMA*, Topics in the Neurosciences, Vol. 9. Springer.

16   European Monitoring Centre for Drugs and Drug Addiction (2019). Ireland Country Drug Report 2019. Available at: http://www.emcdda.europa.eu/countries/drug-reports/2019/ireland_en

17   L.A. Parker (2017). *Cannabinoids and the Brain*. MIT Press.

18   J.L. Cadet et al. (2014). Neuropathology of substance use disorders. *Acta Neuropathologica*, 127: 91–107.

19   RTÉ (2019). 56% of drug addicts abuse prescription drugs, survey suggests, 22 February. Available at: https://www.rte.ie/news/dublin/2019/0222/1032123-drugs

20   National Safety Council (2020). Opioids drive addiction, overdose. Available at: https://www.nsc.org/home-safety/safety-topics/opioids

21   B. Meier (2018). *Pain Killer: An Empire of Deceit and the Origins of America's Opioid Epidemic*. Random House.

22   N. Ohler (2015). *Blitzed: Drugs in the Third Reich*. Kiepenheuer & Witsch.

23   P. Radden Keefe (2017). The family that built an empire of pain. *New Yorker*, 23 October. Available at: https://www.newyorker.com/magazine/2017/10/30/the-family-that-built-an-empire-of-pain

24   A.V. Zee (2009) The promotion and marketing of OxyContin: commercial triumph, public health tragedy. *American Journal of Public Health*, 99: 221–227.

25   Walters, J. (2019). OxyContin maker expected 'a blizzard of prescriptions' following drug's launch. *The Guardian*, 16 January.

26   Rutland Centre (n.d.). Treating gambling addiction. Available at: https://www.rutlandcentre.ie/addictions-we-treat/gambling

27 *Fresh Air* (2019). A neuroscientist explores the biology of addiction in 'never enough'. Interview with Judith Grisel, 12 February. Available at: https://www.npr.org/transcripts/693814827

28 N.D. Volkow and M. Muenke (2012). The genetics of addiction. *Human Genetics*, 131: 773–777.

29 D. Demontis et al. (2019). Genome-wide association study implicates CHRNA2 in cannabis use disorder. *Nature Neuroscience*, 22(7): 1066–1074.

30 C. Pickering et al. (2008). Sensitization to nicotine significantly decreases expression of GABA transporter GAT-1 in the medial prefrontal cortex. *Progress in Neuro-Psychopharmacology and Biological Psychiatry*, 32: 1521–1526.

31 C.N. Simonti et al. (2016). The phenotypic legacy of admixture between modern humans and Neanderthals. *Science*, 351: 737–774.

32 American Psychiatric Association (2013). *Diagnostic and Statistical Manual of Mental Disorders: DSM-5* (5th edition), 490–497.

33 A. Agrawal et al. (2012). The genetics of addiction – a translational perspective. *Translational Psychiatry*, 2:e140.

34 Foundations Recovery Network (2018). Pros and cons of decriminalizing drug addiction, 23 April. Available at: https://www.foundationsrecoverynetwork.com/pros-and-cons-of-decriminalizing-drug-addiction/

35 J.M. Solis et al. (2012). Understanding the diverse needs of children whose parents abuse substances. *Current Drug Abuse Reviews* 5: 135–147.

36 M. Liu et al. (2019). Association studies of up to 1.2 million individuals yield new insights into the genetic etiology of tobacco and alcohol use. *Nature Genetics* 51, 237–244.

37 J. Mennis et al. (2016). Risky substance use environments and addiction: a new frontier for environmental justice research. *International Journal of Environmental Research and Public Health*, 13: 607.

38 M. Enoch (2011). The role of early life stress as a predictor for alcohol and drug dependence. *Psychopharmacology* (Berlin), 214: 17–31.

39 Child maltreatment and alcohol. *World Health Organization*. Available at: https://www.who.int/violence_injury_prevention/violence/world_report/factsheets/fs_child.pdf

40 G.P. Lee et al. (2012). Association between adverse life events and addictive behaviors among male and female adolescents. *American Journal on Addictions*, 516–523.

41 National Institute on Drug Abuse (2016). *Principles of substance abuse prevention for early childhood: a research-based guide*. Chapter 1: Why Is Early Childhood Important to Substance Abuse Prevention?

42 E.G. Spratt et al. (2012) The effects of early neglect on cognitive, language, and behavioral functioning in childhood. *Psychology*, 3: 175–182.

43 A.G.P. Wakeford et al. (2018). A review of non-human primate models of early life stress and adolescent drug abuse. *Neurobiology of Stress*, 9: 188–198.

44 L.I. Sederer (2019). What does 'Rat Park' teach us about addiction? *Psychiatric Times*, 10 June.

45 H. Carliner (2016). Childhood trauma and illicit drug use in adolescence: a population-based national comorbidity survey replication-adolescent supplement study. *Journal of the American Academy of Child and Adolescent Psychiatry*, 55, 701–708.

46 C.J. Hammond (2014). Neurobiology of adolescent substance use and addictive behaviors: prevention and treatment implications. *Adolescent Medicine: State of the Art Reviews*, 25: 15–32.

47 Age and substance abuse. *Alcohol Rehab*. Available at: https://alcoholrehab.com/drug-addiction/age-and-substance-abuse/

48 J.S. Fowler et al. (2007). Imaging the addicted human brain. *Science and Practice Perspectives*, 3: 4–16.

49 M. Ushe and J.S. Perlmutter (2013). Sex, drugs and Parkinson's disease. *Brain*, 136: 371–373.

50 L. Holmes (2018). A reminder that addiction is an illness, not a character flaw. *HuffPost*, 26 July. Available at: https://www.huffpost.com/entry/addiction-stigma-how-to help_n_5b58806ae4b0b15a-ba942161

51 J. Hartmann-Boyce et al. (2018). Nicotine replacement therapy versus control for smoking cessation. *Cochrane Systematic Review*. Available at: https://www.cochranelibrary.com/cdsr/doi/10.1002/14651858.CD000146.pub5/full

52 P. Hajek et al. (2019). A randomized trial of e-cigarettes versus nicotine-replacement therapy. *New England Journal of Medicine*, 380: 629–637.

CHAPTER 7

1 United Nations Office on Drugs and Crime (2019). World Drug Report 2019. Available at: https://wdr.unodc.org/wdr2019/

2 Monarch Shores. How much does the war on drugs cost? Available at: https://www.monarchshores.com/drug-addiction/how-much-does-the-war-on-drugs-cost

3 Drug Policy Alliance (n.d.). Drug war statistics. Available at: http://www.drugpolicy.org/issues/drug-war-statistics

4 G. Borsa (2019). Drug markets in Europe estimated to be worth $30 billion. A thriving market that empowers organized crime, posing a threat to society as a whole. *SIR: Agenzia d'Informatizone*, 26 November.

5 C.J. Coyne and A.R. Hall (2017). Four decades and counting: the continued failure of the war on drugs. CATO Institute report, 12 April.

6 A. Lockie (2019). Top Nixon adviser reveals the racist reason he started the 'war on drugs' decades ago. *Business Insider*, 31 July. Available at: https://www.businessinsider.com/nixon-adviser-ehrlichman-anti-left-anti-black-war-on-drugs-2019-7?r=US&IR=T

7   W.H. Park (1898). *Opinions of Over 100 Physicians on the Use of Opium in China.*

8   E. Trickey (2018). Inside the story of America's 19th-century opiate addiction. *Smithsonian Magazine*, 4 January.

9   E. Brecher et al. (1972). The Consumers Union report on licit and illicit drugs. UK Cannabis Internet Activist. Available at: https://www.ukcia.org/research/cunion/cu6.htm

10 B. Fairy (n.d.). How marijuana became illegal. Available at: http://www.ozarkia.net/bill/pot/blunderof37.html

11 United States Congress Senate Committee on the Judiciary (1955). *Communist China and illicit narcotic traffic.* United States Government Printing Office.

12 J. Clear (2018). *Atomic Habits: An Easy & Proven Way to Build Good Habits & Break Bad Ones.* Penguin Random House.

13 S.X. Zhang and K.L. Chin (2016). A people's war: China's struggle to contain its illicit drug problem. *Foreign Policy at Brookings.* Available at: https://www.brookings.edu/wp-content/uploads/2016/07/A-Peoples-War-final.pdf

14 M.A. Lee and B. Shlain (1992). *Acid Dreams: The Complete Social History of LSD: The CIA, the Sixties, and Beyond.* Grove Press.

15 M.P. Bogenschutz and S. Ross (2018). Therapeutic applications of classic hallucinogens. *Current Topics in Behavioral Neuroscience*, 36: 361–391.

16 US Congressional Record. Controlled Substances Act. Available at: https://www.congress.gov/congressional-record/congressional-record-index/114th-congress/1st-session/controlled-substances-act/7918

17 Misuse of Drugs (Amendment) Act 2015. Available at: http://www.irishstatutebook.ie/eli/2015/act/6/enacted/en/html

18 F. Schifano (2018). Recent changes in drug abuse scenarios: the new/novel psychoactive substances (NPS) phenomenon. *Brain. Sciences*, 8(12): 221.

19 K. Holland (2018). Almost 75% of drugs offences last year were 'possession for personal use'. *Irish Times*, 24 June.

20 E. Dufton (2017). *Grass Roots: The Rise and Fall and Rise of Marijuana in America.* Hachette.

21 Sentencing Project (2018). Report to the United Nations on Racial Disparities in the U.S. Criminal Justice System. Available at: https://www.sentencingproject.org/publications/un-report-on-racial-disparities/

22 American Civil Liberties Union (2020). Marijuana arrests by the numbers. Available at: https://www.aclu.org/gallery/marijuana-arrests-numbers

23 Jenny Gesley (2016). *Decriminalization of Narcotics: Netherlands.* Library of Congress, July.

24 A. Bell (1999). Deaths soar as Dutch drugs flood in. *The Observer*, 5 September.

25 C. Ort et al. (2014). Spatial differences and temporal changes in illicit drug use in Europe quantified by wastewater analysis, *Addiction* 109: 1338–1352.

26 A. Ritter et al. (2013). Government drug policy expenditure in Australia 2009–2010. *Drug Policy Modelling Program Monograph Series.* Sydney: National Drug and Alcohol Research Centre.

27 T. Makkai et al. (2018). Report on Canberra GTM Harm Reduction Service. Available at: https://www.harmreductionaustralia.org.au/wp-content/uploads/2018/06/Pill-Testing-Pilot-ACT-June-2018-Final-Report.pdf

28 Australian Criminal Intelligence Commission (2018). Illicit Drug Data Report 2016–17. Available at: https://www.acic.gov.au/publications/intelligence-products/illicit-drug-data-report-2016-17

29 Drug Policy Alliance (2019). Drug decriminalization in Portugal: learning from a health- and human-centered approach. Available at: http://www.drugpolicy.org/resource/drug-decriminalization-portugal-learning-health-and-human-centered-approach

30 N. Bajekal (2018). Want to win the war on drugs? Portugal might have the answer. *Time*, 1 August.

31 D.J. Nutt et al (2010). Drug harms in the UK: a multicriteria decision analysis. *Lancet.* 376 1558–1565

32 Foundations Recovery Network (2018). Pros and cons of decriminalizing drug addiction, 23 April. Available at: https://www.foundationsrecoverynetwork.com/pros-and-cons-of-decriminalizing-drug-addiction/

33 Partnership for Drug-Free Kids. Preventing teen drug use: risk factors & why teens use. Available at: https://drugfree.org/article/risk-factors-why-teens-use

34 L.M. Squeglia et al. (2009). The influence of substance use on adolescent brain development. *Clinical EEG and Neuroscience*, 40: 31–38.

35 Substance Abuse and Mental Health Services Administration (USA) (1999). *Treatment Improvement Protocol (TIP) Series*, No. 33. Chapter 2: How Stimulants Affect the Brain and Behavior. Available at: https://www.ncbi.nlm.nih.gov/books/NBK64328/

36 F. Muller et al. (2018). Neuroimaging of chronic MDMA ('ecstasy') effects: a meta-analysis. *Neuroscience and Biobehavioral Reviews*, 96: 10–20.

37 M.D. Wunderli et al. (2018). Social cognition and interaction in chronic users of 3,4-Methylenedioxymethamphetamine (MDMA, 'Ecstasy'). *International Journal of Neuropsychopharmacology*, 21: 333–344.

38 G. Gobbi et al. (2019). Association of cannabis use in adolescence and risk of depression, anxiety and suicidality in young adulthood: a systematic review and meta-analysis. *JAMA Psychiatry*, 76: 426–434.

39 C.L. Odgers (2008). Is it important to prevent early exposure to drugs and alcohol among adolescents? *Psychological Science*, 19, 1037–1044.

40 A. Jaffe (2018). Is marijuana a gateway drug? *Psychology Today*, 24 July.

41   D.M. Anderson (2019). Association of marijuana laws with teen marijuana use: new estimates from the youth risk behavior surveys. *JAMA Pediatrics*, 173: 879–881.

42   M. Cerda et al. (2019). Association between recreational marijuana legalization in the United States and changes in marijuana use and cannabis use disorder from 2008 to 2016. *JAMA Psychiatry*, 13 November. doi:10.1001/jamapsychiatry.2019.3254.

43   F. Tennant (2013). Elvis Presley: Head trauma, autoimmunity, pain, and early death. *Practical Pain Management*, 13(5).

44   K. Harmon (2011). What is Propofol and how could it have killed Michael Jackson? *Scientific American*, 3 October.

45   M. Puente (2018). Prince's death: Superstar didn't know he was taking fentanyl; no one charged with a crime. *USA Today*, 19 April.

46   A. Topping (2013). Amy Winehouse died of alcohol poisoning, second inquest confirms. *The Guardian*, 8 January.

47   J. Elflein (2019). Drug use in the U.S. – statistics and facts. Statista, 10 September. Available at: https://www.statista.com/topics/3088/drug-use-in-the-us/

CHAPTER 8

1   M. Hamer (1990). No forensic evidence against Birmingham Six. *New Scientist*, 24 November.

2   Alpha-1 Foundation Ireland (n.d.). New study shows health benefits of the smoking ban in Ireland. Available at: https://www.alpha1.ie/news-events/latest-news/149-new-study-shows-the-smoking-ban-improves-health

3   J.K. Hamlin and K. Wynn (2011). Young infants prefer prosocial to antisocial others. *Cognitive Development*, 26: 30–39.

4   G. Carra (2004). Images in psychiatry: Cesare Lombroso, M.D. 1835–1909. *American Journal of Psychiatry*, 161: 624

5   G.F. Vito, J.R. Maahs and R.M. Holmes (2007). *Criminology: Theory, Research, and Policy*. Jones and Bartlett.

6   L. Moccia (2018). The Experience of Pleasure: A perspective between neuroscience and psychoanalysis. *Fronrs in Human Neuroscience*, 12: 359.

7   Population Reference Bureau (2012). U.S. has world's highest incarceration rate. Available at: https://www.prb.org/us-incarceration/

8   Irish Penal Reform Trust (2020). Facts & figures. Available at: https://www.iprt.ie/prison-facts-2/

9   Statista (2017). The prison gender gap. Available at: https://www.statista.com/chart/11573/gender-of-inmates-in-us-federal-prisons-and-general-population/

10   S. Kang (2014). Why do young men commit more crimes? *Economics of Crime* online course, Hanyang University. FutureLearn.

11   UNODC Global Study on Homicide 2013: Trends, Context, Data (2013). UNODC. Available at: https://www.unodc.org/documents/data-and-analysis/statistics/GSH2013/2014_GLOBAL_HOMICIDE_BOOK_web.pdf

12   H.J. Janssen et al. (2017). Sex differences in longitudinal pathways from parenting to delinquency. *European Journal on Criminal Policy and Research*, 23: 503–521.

13   Lexercise (n.d.) Do more boys than girls have learning disabilities? Available at: https://www.lexercise.com/blog/boys-girls-learning-disabilities

14   M.L. Batrinos (2012). Testosterone and aggressive behavior in man. *International Journal of Endocrinology and Metabolism*, 10: 563–568.

15   D. Hollman and E. Alderman (2008). Fatherhood in adolescence. *Pediatrics in Review*, 29: 364–366.

16   S. Scheff (2017). More boys admit to cyberbullying than girls. *Psychology Today*, 5 October.

17   B. Bell (2015). Do recessions increase crime? World Economic Forum report, 4 March.

18   A. Burke and D. Chadee (2018). Effects of punishment, social norms, and peer pressure on delinquency: spare the rod and spoil the child? *Journal of Social and Personal Relationships* 36(9): 2714–2737.

19   A. Raine (2014). *The Anatomy of Violence: The Biological Roots of Crime*. Vintage Books.

20   K.O. Christiansen (1970). Crime in a Danish twin population. *Acta Geneticae Medicae et Gemellologiae* (Roma), 19: 323–326.

21   R.R. Crowe (1972). The adopted offspring of women criminal offenders. A study of their arrest records. *Archives of General Psychiatry*, 27: 600–603.

22   M. Bohman (1978). Some genetic aspects of alcoholism and criminality. *Archives of General Psychiatry*. 35, 269–276.

23   S.A. Mednick, W.F. Gabrielli and B. Hutchings (1983). *Genetic Influences in Criminal Behavior – Evidence from an Adoption Cohort. Prospective Studies on Crime and Deliquency*. Kluwer-Nijhoff.

24   S. Sohrabi (2015). The criminal gene: the link between MAOA and aggression. *BMC Proceedings*, 9 (Suppl. 1): A49.

25   H.G. Brunner (1993). Abnormal behavior associated with a point mutation in the structural gene for monoamine oxidase A. *Science*, 262: 578–80.

26   V. Nikulina, C. Spatz Widom and L.M. Brzustowicz (2012). Child abuse and neglect, MAO-A, and mental health outcomes: a prospective examination. *Biological Psychiatry*, 71: 350–357.

27   E. Salinsky (2018). Violence is preventable. *Grantmakers in Health*, March.

28   Central Statistics Office (2019). Recorded crime victims 2018. Available at: https://www.cso.ie/en/releasesandpublications/ep/p-rcv/recordedcrimevictims2018/

29   D.A. Stetler et al. (2014). Association of low-activity MAOA allelic variants with violent crime in incarcerated offenders. *Journal of Psychiatric Research*, 58: 69–75.

30 S.C. Godar et al. (2011). Maladaptive defensive behaviours in monoamine oxidase A-deficient mice.
International Journal of Neuropsychopharmacology,
14: 1195–1207.

31 Y. Kuepper et al. (2013). MAOA-uVNTR genotype
predicts interindividual differences in experimental aggressiveness as a function of the degree of
provocation. *Behavioural Brain Research*, 247:73–78.

32 L.M. Williams (2009). A polymorphism of
the MAOA gene is associated with emotional brain
markers and personality traits on an antisocial
index. *Neuropsychopharmacology*, 34: 1797–1809.

33 D.M. Fergusson et al. (2011). MAO-A, abuse exposure and antisocial behaviour: 30-year longitudinal
study. *British Journal of Psychiatry*, 198: 457–463.

34 T.K. Newman et al. (2005). Monoamine oxidase A
gene promoter variation and rearing experience
influences aggressive behavior in rhesus monkeys.
Biological Psychiatry, 15 January; 57(2): 167–172.

35 F.E.A. Verhoeven et al. (2012). The effects of
MAOA genotype, childhood trauma, and sex on
trait and state-dependent aggression. *Brain and
Behavior*, 2: 806–813.

36 S. McSwiggan, B. Elger and P.S. Appelbaum (2017).
The forensic use of behavioral genetics in criminal
proceedings: case of the MAOA-L genotype. *International Journal of Law and Psychiatry*, 50: 17–23.

37 M.L. Baum (2009). The Monoamine Oxidase A
(MAOA) genetic predisposition to impulsive violence: is it relevant to criminal trials? *Neuroethics*,
doi 10.1007/s12152-011-9108-6.

38 D.A. Crighton and G.J. Towl (2015). *Forensic Psychology*. John Wiley & Sons.

39 V.A. Toshchakova et al. (2018). Association of
polymorphisms of serotonin transporter (5HT-
TLPR) and 5-HT2C receptor genes with criminal
behavior in Russian criminal offenders. *Neuropsy-
chobiology*, 75(4): 200–210.

40 N. Larsson (2015). 24 ways to reduce crime in the
world's most violent cities. *The Guardian*, 30 June.

41 C. O'Keefe (2019). Sharp rise in violent crime
since last year. *Irish Examiner*, 26 March.

CHAPTER 9

1 Intersex Society of North America (n.d.). How
common is intersex? Available at: https://isna.org/
faq/frequency/

2 A. Alvergne (2016). Do women's periods really
synch when they spend time together? *The Conversation*, 14 July. Available at: https://theconversation.
com/do-womens-periods-really-synch-when-they-
spend-time-together-61890

3 Usable Stats (n.d.). Fundamentals of statistics 2:
the normal distribution. Available at: https://www.
usablestats.com/lessons/normal

4 La Griffe du Lion (2000). Aggressiveness, criminality and sex drive by race, gender and ethnicity, 2(11).
Accessible at: http://lagriffedulion.f2s.com/fuzzy.htm

5 S.T. Ngo et al. (2014). Gender differences in

autoimmune disease. *Frontiers in Neuroendocrinolo-
gy*, 35 (3): 347–69. doi:10.1016/j.yfrne.2014.04.004

6 T.M. Wizemann and M.L. Pardue (2001). *Commit-
tee on Understanding the Biology of Sex and Gender
Differences: Exploring the Biological Contributions to
Human Health: Does Sex Matter?* National Academy
Press.

7 Harvard Men's Health Watch (2019). Mars vs. Venus: the gender gap in health. Available at: https://
www.health.harvard.edu/newsletter_article/mars-
vs-venus-the-gender-gap-in-health

8 G. Lawton (2020). Why are men more likely to get
worse symptoms and die from COVID-19? *New
Scientist*, 16 April.

9 O. Ryan (2018). Men account for eight in ten suicides in Ireland. *The Journal*,4 October. Available
at: https://www.thejournal.ie/suicide-rates-ireland-
4267893-Oct2018/

10 S. Naqvi et al. (2019). Conservation, acquisition,
and functional impact of sex-biased gene expression in mammals. *Science*, 365(6450) pii: eaaw7317.

11 S. McDermott (2019). Life expectancy: Gap
narrows between Irish men and women (but both
are living longer than before). *The Journal*, 23
December. Available at: www.thejournal.ie/life-
expectancy-ireland-2019-4947423-Dec-2019

12 S.N. Austad (2006). Why women live longer than
men: sex difference in longevity. *Gender Medicine*
3(2): 79–92.

13 M. Roser et al. (2019). Life expectancy. Our World
in Data. Available at: https://ourworlddata.org/
why-do-women-live-longer-than-men

14 S.H. Preston and H. Wang (2006). Sex mortality
differences in the United States? The role of cohort smoking patterns. *Demography* 43(4): 631–646.

15 D. Iliescu et al. (2016). Sex differences in intelligence: A multi-measure approach using nationally
representative samples from Romania. *Intelligence*,
58: 54–61.

16 J. Shibley Hyde (2005). The gender similarities
hypothesis. *American Psychologist*, Vol. 60: 581–592.

17 A. Grant (2019). Differences between men
and women are vastly exaggerated. *Human
Resources*, 1 July. Available at: https://www.
humanresourcesonline.net/differences-be-
tween-men-and-women-are-vastly-exaggerated

18 T. Kaiser et al. (2019). Global sex differences in
personality: replication with an open online data-
set. *Journal of Personality*, 2019; 00: 1–15.

19 C. Fine (2010). *Delusions of Gender: How Our
Minds, Society and Neurosexism Create Difference*.
W.W. Norton.

20 D. Joseph and D.A. Newman (2010). Emotional
intelligence: an integrative meta-analysis and cas-
cading model. *Journal of Applied Psychology*, 95(1):
54–78. doi:10.1037/a0017286.

21 D. Goleman (2011). Are women more emotionally
intelligent than men? *Psychology Today*, 20 April.

22 C.V. Mitchell and R. Koonce (2019). Leadership

traits that transcend gender. *Chief Learning Officer*, 10 June. Available at: https://www.chieflearningofficer.com/2019/06/10/leadership-traits-that-transcend-gender/

23 F. de Waal (2019). What animals can teach us about politics. *The Guardian*, 12 March.

24 Z. Mejia (2018). Just 24 female CEOs lead the companies on the 2018 Fortune 500 – fewer than last year. CNBC *Make It*, 21 May. Available at: https://www.cnbc.com/2018/05/21/2018s-fortune-500-companies-have-just-24-female-ceos.html

25 S. Gausepohl (2016). 3 steps women can take to blaze a leadership trail. *Business Daily News*, 15 December.

26 M. Staines (2019). Survey warns women in senior roles more likely to face discrimination at work than men. Newstalk, 13 September. Available at: https://www.newstalk.com/news/women-discrimination-workplace-904130

27 All Diversity (n.d.) 17 reasons women make great leaders. Available at: https://alldiversity.com/news/17-Reasons-Women-Make-Great-Leaders

28 M. Rosencrans (2019). Women make great leaders: four ways to embrace and advance your leadership skills. *Forbes Communications Council*, 4 March.

29 R. Riffkin (2014). Americans still prefer a male boss to a female boss. Gallup Poll (Economics), 14 October.

30 R.J. Haier et al. (2005). The neuroanatomy of general intelligence: sex matters. *Neuroimage*, 25(1): 320–327.

31 L. Eliot (2019). Bad science and the unisex brain. *Nature*, 566: 454–455.

32 D.F. Swaab and E. Fliers (1985). A sexually dimorphic nucleus in the human brain. *Science*, 228, 1112–1115.

33 M. Price (2017). Study finds some significant differences in brains of men and women. *Science*, 11 April doi:10.1126/science.aal1025.

34 M. Ingalhalikar et al. (2014). Sex differences in the structural connectome of the human brain. *Proceedings of the National Academy of Sciences*, 111(2): 823–828.

35 M.M. Lauzen et al. (2008). Constructing gender stereotypes through social roles in prime-time television. *Journal of Broadcasting & Electronic Media*, 52: 200–214.

36 J. McCabe et al. (2011). Gender in twentieth-century children's books: patterns of disparity in titles and central characters. *Gender and Society*, 25: 197–226.

37 OECD (2015). *The ABC of Gender Equality in Education: Aptitude, Behaviour, Confidence.* OECD. Available at: https://www.oecd.org/pisa/keyfindings/pisa-2012-results-gender-eng.pdf

38 V. LoBue (2019). Are boys really better than girls at math and science? *Psychology Today*, 8 April.

39 C. O'Brien (2019). Girls outperform boys in most Leaving Cert subjects at higher level. *Irish Times*,

15 August.

40 S. Kuper and E. Jacobs (2019). The untold danger of boys falling behind in school. *Daily Dose*, 13 January. Available at: https://www.ozy.com/fast-forward/the-untold-danger-of-boys-falling-behind-in-school/91361

41 J. McCurry (2018). Two more Japanese medical schools admit discriminating against women. *The Guardian*, 12 December.

42 J. Marcus (2017). Why men are the new college minority. *The Atlantic*, 8 August. Available at: https://www.theatlantic.com/education/archive/2017/08/why-men-are-the-new-college-minority/536103/

43 A. Harris (2018). The problem with all-girls' schools. *Irish Times*, 27 February.

44 O. James (2009). Family under the microscope: the alarming rate of distress among 15-year-old girls affects all classes. *The Guardian*, 25 July.

45 E. Smyth (2010). Single-sex education: what does research tell us? *Revue Française de Pédagogie*, 171: 47–55.

46 G. Hamman (2013). German government campaigns for more male kindergarten teachers. *DW*, 8 October. Available at: https://www.dw.com/en/german-government-campaigns-for-more-male-kindergarten-teachers/a-17143449

47 M.J. Perry (2018). Chart of the day: the declining female share of computer science degrees from 28% to 18%. American Enterprise Institute, 6 December. Available at: https://www.aei.org/carpe-diem/chart-of-the-day-the-declining-female-share-of-computer-science-degrees-from-28-to-18

48 OECD (2017). Women make up most of the health sector workers but they are under-represented in high-skilled jobs. OECD, March. Available at: https://www.oecd.org/gender/data/women-make-up-most-of-the-health-sector-workers-but-they-are-under-represented-in-high-skilled-jobs.htm

49 Eurostat Press Office (2018). Women in the EU earned on average 16% less than men in 2016. Eurostat News Release, 8 March. Available at: https://ec.europa.eu/eurostat/documents/2995521/8718272/3-07032018-BP-EN.pdf/fb402341-e7fd-42b8-a7cc-4e33587d79aa

50 J. Doward and T. Fraser (2019). Hollywood's gender pay gap revealed: male stars earn $1m more per film than women. *The Guardian*, 15 September.

51 YouGov (2013). Women do all the work this Christmas. YouGov, 22 December. Available at: https://yougov.co.uk/topics/politics/articles-reports/2013/12/22/women-do-all-the-work-christmas

52 R. Jensen and E. Oster (2009). The power of TV: cable television and women's status in India. *Quarterly Journal of Economics*, 124: 1057–1094.

53 AFP (Paris) (2019). French toymakers sign pact to rid games and toys of gender stereotypes. *The Journal*, 24 September. Available at: https://www.thejournal.ie/gender-stereotypes-toys-france-4823655-Sep2019/

54 S. Murray (2020). Two lads from Cork have won this year's BT Young Scientists top award. *The Journal*, 10 January. Available at: https://www.the-journal.ie/young-scientist-2020-4961513-Jan2020/

55 K. Langin (2018). What does a scientist look like? Children are drawing women more than ever before. *Science*, 20 March.

56 L.W. Wilde (1997) *Celtic Women in Legend, Myth and History*. Sterling Publishing Co.

CHAPTER 10

1 C. Stringer and J. Galway-Witham (2018). When did modern humans leave Africa? *Science*, 359: 389–390.

2 J. Gabbatiss (2017). Nasty, brutish and short: are humans DNA-wired to kill? *Scientific American*, 19 July.

3 F. Marlowe (2010). *The Hadza Hunter-Gatherers of Tanzania*. University of California Press.

4 S. Müller-Wille (2014). *Linnaeus and the Four Corners of the World: The Cultural Politics of Blood, 1500–1900*. Palgrave Macmillan.

5 R. Bhopal (2007). The beautiful skull and Blumenbach's errors. *British Medical Journal*, 22–29 December.

6 W.H. Goodenough (2002). Anthropology in the 20th century and beyond. *American Anthropologist*, 104: 423–440.

7 T. Ott (2019). How Jesse Owens foiled Hitler's plans for the 1936 Olympics. *Biography*, 20 June.

8 E. Kolbert (2018). There's no scientific basis for race – it's a made-up label. *National Geographic*. Available at: https://www.nationalgeographic.com/magazine/2018/04/race-genetics-science-africa/

9 A.R. Templeton (2019). *Human Population Genetics and Genomics*. Academic Press.

10 A. Gibbons (2015). How Europeans evolved white skin. *Science*, 2 April. doi: 10.1126/science.aab2435.

11 R.P. Stokowski et al. (2007). A genome-wide association study of skin pigmentation in a South Asian population. *American Journal of Human Genetics*, 81: 1119–1132.

12 J.K. Wagner (2017). Anthropologists' views on race, ancestry, and genetics. *American Journal of Physical Anthropology*, 162: 318–327.

13 A. Arenge et al. (2018). Poll: 64 percent of Americans say racism remains a major problem. NBC News, 29 May. Available at: https://www.nbcnews.com/politics/politics-news/poll-64-per cent-americans-say-racism-remains-major-problem-n877536

14 A. Brown (2019). Key findings on Americans' views of race in 2019. Pew Research Centre, 9 April. Available at: https://www.pewresearch.org/fact-tank/2019/04/09/key-findings-on-americans-views-of-race-in-2019/

15 M. Snow (2019). Trump is racist, half of US voters say. Quinnipiac University Poll. Available at: https://poll.qu.edu/national/release-detail?ReleaseID=3636

16 G. Armstrong (2003). *Football Hooligans: Knowing the Score*. Explorations in Anthropology. Berg.

17 D. Kilvington (2019). Racist abuse at football games is increasing, Home Office says – but the sport's race problem goes much deeper. *The Conversation*, 9 October. Available at: https://theconversation.com/racist-abuse-at-football-games-is-increasing-home-office-says-but-the-sports-race-problem-goes-much-deeper-124467

18 Bridge Initiative Team (2019). Factsheet: polls on Islam, Muslims and Islamophobia in Canada. Georgetown University. Available at: https://bridge.georgetown.edu/research/fact-sheet-polls-on-islam-muslims-and-islamophobia-in-canada/

19 R. Reeve (2015). *Infamy: The Shocking Story of the Japanese American Internment in World War II*. Henry Holt & Co.

20 S. Yamoto (1997). *Personal Justice Denied. Report of the Commission on Wartime Relocation and Internment of Civilians*. University of Washington Press.

21 Pew Research Center (2011). Muslim-Western tensions persist: common concerns about Islamic extremism. Pew Research Center. Available at: https://www.pewresearch.org/global/2011/07/21/muslim-western-tensions-persist/

22 Human Rights Watch (2012). World Report 2012: Israel/Occupied Palestinian Territories: Events of 2011. Human Rights Watch. Available at: https://www.hrw.org/world-report/2012/country-chapters/israel/palestine

23 M.G. Bard (2020). Human rights in Israel: background and overview. Jewish Virtual Library. Available at: https://www.jewishvirtuallibrary.org/background-and-overview-of-human-rights-in-israel

24 T. Stafford (2017). This map shows what white Europeans associate with race – and it makes for uncomfortable reading. *The Conversation*, 2 May. Available at: https://theconversation.com/this-map-shows-what-white-europeans-associate-with-race-and-it-makes-for-uncomfortable-reading-76661

25 World Bank et al. (2018). Overcoming Poverty and Inequality in South Africa. International Bank for Reconstruction and Development, and World Bank. Available at: http://documents.worldbank.org/curated/en/530481521735906534/pdf/124521-REV-OUO-South-Africa-Poverty-and-Inequality-Assessment-Report-2018-FINAL-WEB.pdf

26 M. O'Halloran (2019). Ireland has 'worrying pattern' of racism, head of EU agency warns. *Irish Times*, 27 September.

27 Department of Justice and Equality (2017). *National Traveller and Roma Inclusion Strategy 2017–2021*. Department of Justice and Equality. Available at: http://www.justice.ie/en/JELR/National%20Traveller%20and%20Roma%20Inclusion%20Strategy,%202017-2021.pdf/Files/National%20Traveller%20and%20Roma%20Inclusion%20Strategy,%202017-2021.pdf

28   J. O'Connell (2013). Our casual racism against Travellers is one of Ireland's last great shames. *Irish Times*, 27 February.

29   IrishHealth.com (n.d.). Health and the Travelling community. IrishHealth.com. Available at: http://www.irishhealth.com/article.html?id=1079

30   B. Shoot (2019). Immigrants founded nearly half of 2018's *Fortune* 100 companies, new data analysis shows. *Fortune*, 15 January.

31   D. Kosten (2018). Immigrants as economic contributors: immigrant entrepreneurs. National Immigration Forum. 11 July. Available at: https://immigrationforum.org/article/immigrants-as-economic-contributors-immigrant-entrepreneurs/

32   T. Jawetz (2019). Building a more dynamic economy: The benefits of immigration. Testimony before the US House Committee on the Budget. Centre for American Progress. Available at: https://docs.house.gov/meetings/BU/BU00/20190626/109700/HHRG-116-BU00-Wstate-JawetzT-20190626.pdf

33   The Sentencing Project (2019). Criminal justice facts. Sentencing Project. Available at: https://www.sentencingproject.org/criminal-justice-facts/

34   C. Kenny (2017). The data is in: Young people are increasingly less racist than old people. *Quartz*, 24 May. Available at: https://qz.com/983016/the-data-are-in-young-people-are-definitely-less-racist-than-old-people/

35   Reni Eddo-Lodge, Guardian, 30 May 2017 https://www.theguardian.com/world/2017/may/30/why-im-no-longer-talking-to-white-people-about-race

CHAPTER 11

1   AO Show (2018). 83% of Irish workers think about quitting their job every day. iRadio, 20 November. Available at: https://www.iradio.ie/jobdone/

2   K. Iwamoto (2017). East Asian workers remarkably disengaged. *Nikkei Review*, 25 May.

3   A. Adkins (2015). Majority of US employees not engaged despite gains in 2014. Gallup, 28 January. Available at: https://news.gallup.com/poll/181289/majority-employees-not-engaged-despite-gains-2014.aspx

4   D. Spiegel (2019). 85% of American workers are happy with their jobs, national survey shows. CNBC, 2 April. Available at: https://www.cnbc.com/2019/04/01/85percent-of-us-workers-are-happy-with-their-jobs-national-survey-shows.html

5   B. Rigoni and B. Nelson (2016). Few millennials are engaged at work. Gallup, 30 August. Available at: https://news.gallup.com/businessjournal/195209/few-millennials-engaged-work.aspx

6   T. Kohler et al (2017). Greater post-Neolithic wealth disparities in Eurasia than in North America and Mesoamerica. *Nature*, 551: 619–622.

7   *The Economist* (2019). Redesigning the corporate office. *The Economist*, 28 September. Available at: https://www.economist.com/business/2019/09/28/redesigning-the-corporate-office

8   M. Guta (2018). 68 per cent of workers still get most work done in traditional offices. *Small Business Trends*, 13 June.

9   K2Space (n.d.). The history of office design. K2Space. Available at: https://k2space.co.uk/knowledge/history-of-office-design

10   Goldman Sachs (2019). A new European headquarters for Goldman Sachs. Goldman Sachs. Available at: https://www.goldmansachs.com/careers/blog/posts/goldman-sachs-london-plumtree-court.html

11   British Council for Offices (2018). The rise of flexible workspace in the corporate sector. Available at: http://www.bco.org.uk/Research/Publications/The_Rise_of_Flexible_Workspace_in_the_Corporate_Sector.aspx

12   S. Bevan and S. Hayday (2001). Costing Sickness Absence in the UK. Institute for Employment Studies. Available at: https://www.employment-studies.co.uk/system/files/resources/files/382.pdf

13   Unilever (n.d.). Improving employee health & well-being. Unilever. Available at: https://www.unilever.com/sustainable-living/enhancing-livelihoods/fairness-in-the-workplace/improving-employee-health-nutrition-and-well-being/

14   S. Bean (2016). Two-thirds of British workers more productive working in the office. *Insight*, 26 October. Available at: https://workplaceinsight.net/two-thirds-of-british-workers-more-productive-working-in-the-office/

15   J. Oates (2019). Hot desk hell: staff spend two weeks a year looking for seats in open-plan offices *Register*, 21 June. Available at: https://www.theregister.co.uk/2019/06/21/staff_hot_desk_seats/

16   Gallup (2017). State of the Global Workplace. Gallup. Available at: https://www.gallup.de/183833/state-the-global-workplace.aspx

17   J. Butler (2018). Link between earnings and happiness is a tenuous one. *Financial Times*, 7 February.

18   S. Nasiripour (2016). White House predicts robots may take over many jobs that pay $20 per hour. *HuffPost*, 24 June. Available at: https://www.huffpost.com/entry/white-house-robot-workers_n_56cdd89ce4b0928f5a6de955

19   U. Gentilini et al. (2020). *Exploring Universal Basic Income: A Guide to Navigating Concepts, Evidence, and Practices*. World Bank Group.

20   IGM Economic Experts Panel. Universal Basic Income. IGM Forum. Available at: http://www.igmchicago.org/surveys/universal-basic-income/

21   A. Kauranen (2019). Finland's basic income trial boosts happiness but not employment. Reuters, 8 February. Available at: https://www.reuters.com/article/us-finland-basic-income/finlands-basic-income-trial-boosts-happiness-but-not-employment-idUSKCN1PX0NM

22   A.H. Maslow (1943). A theory of human motivation. *Psychological Review*, 50(4): 370–396.

23   J. Gabay (2015). *Brand Psychology: Consumer Perceptions, Corporate Reputations*. Kogan Page.

24 R.T. Kreutzer and K.H. Land (2013). *Digital Darwinism: Branding and Business Models in Jeopardy*. Springer Publishing.

25 D. Graeber (2018). *Bullshit Jobs: A Theory*. Simon & Schuster.

26 S. Cook (2019). *Making a Success of Managing and Working Remotely*. IT Governance Publishing.

27 R. Biederman et al. (2018). *Reimagining Work: Strategies to Disrupt Talent, Lead Change, and Win with a Flexible Workforce*. Wiley.

28 S. Russell (2019). How remote working can increase stress and reduce well-being. *The Conversation*, 11 October. Available at: http://theconversation.com/how-remote-working-can-increase-stress-and-reduce-well-being-125021

29 J. Grenny and D. Maxfield (2017). A study of 1,100 employees found that remote workers feel shunned and left out. *Harvard Business Review*, 2 November.

30 J. Holmes and M. Stubbe (2014). *Power and Politeness in the Workplace*. Routledge.

31 S. Pinker (2015). *The Village Effect: How Face-to-Face Contact can Make Us Healthier and Happier*. Vintage Canada.

32 P. Gustavson and S. Liff (2014). *A Team of Leaders: Empowering Every Member to Take Ownership, Demonstrate Initiative and Deliver Results*. American Management Association.

33 A. Grant (2011). How customers can rally your troops. *Harvard Business Review*, June.

34 H.P. Gunz and M. Peiperi (2007). *Handbook of Career Studies*. Sage.

35 L. Kellaway (2012). Manual work holds the key to spiritual bliss. *Financial Times*, 3 June.

36 B. Mitchell (2017). Unemployment is miserable and doesn't spawn an upsurge in personal creativity. Bill Mitchell – Modern Monetary Theory, 21 November. Available at: http://bilbo.economicoutlook.net/blog/?p=37429

CHAPTER 12

1 Credit Suisse (2019). Global Wealth Report 2019. Available at: https://www.credit-suisse.com/about-us/en/reports-research/global-wealth-report.html

2 M. Goldring (2017). Eight men own more than 3.6 billion people do: our economics is broken. *The Guardian*, 16 January.

3 Wikipedia (n.d.). Distribution of wealth. Available at: https://en.wikipedia.org/wiki/Distribution_of_wealth

4 Wealth-X Billionaire Census 2019. Available at: https://www.wealthx.com/report/the-wealth-x-billionaire-census-2019/

5 E. Horton (2019). Female billionaires: the new emerging growth market. *Financial News*. 8 November.

6 P. Jacobs (2014). The 20 universities that have produced the most billionaires. *Business Insider Australia*, 18 September. Available at: https://www.businessinsider.com.au/universities-with-most-billionaire-undergraduate-alumni-2014-9

7 L. Stangel (2018). Stanford mints more billionaires than any other college on the planet – except one. Silicon Valley. *Business Journal*, 18 May.

8 *Forbes* (n.d.). World's billionaires list. Available at: https://www.forbes.com/billionaires/#2db-1b704251c

9 F. Reddan (2019). Number of Irish millionaires rises by 3,000 to nearly 78,000. *Irish Times*, 13 March.

10 C. Clifford (2016). 62 percent of American billionaires are self-made. *Entrepreneur Europe*. 14 January.

11 M. Henney (2019). How much do billionaires donate to charity? *Fox Business*, 26 November.

12 Wealth-X Philanthropy Report 2019. Available at: https://www.wealthx.com/report/uhnw-giving-philanthropy-report-2019/

13 Giving Pledge (website). https://givingpledge.org/

14 Donald Read (1994). *The Age of Urban Democracy: England 1868–1914*. Routledge.

15 P. Malpass (1998). *Housing, Philanthropy and the State: A History of the Guinness Trust*. University of the West of England

16 Iveagh Trust (website). http://www.theiveaghtrust.ie/?m=2018

17 National Philanthropic Trust (2019). The 2019 DAF Report. Available at: https://www.nptrust.org/philanthropic-resources/charitable-giving-statistics/

18 O. Ryan (2018). Irish charities have an annual income of €14.5 billion and employ 189,000 people. *The Journal*, 25 July. Available at: https://www.thejournal.ie/irish-charities-4145144-Jul2018/

19 H. Waleson (2017). *Atlantic Insights: Giving While Living*. Atlantic Philanthropies.

20 D. Russakoff (2015). *The Prize: Who's in Charge of America's Schools?* Houghton Mifflin Harcourt.

21 C. Fiennes (2017). We need a science of philanthropy. Nature, 546: 187.

22 Center for Effective Philanthropy. Philanthropy Awards, 2017. Available at: https://cep.org/2017-in-the-news/

23 Bill & Melinda Gates Foundation (n.d.) Who we are: foundation fact sheet. Available at: https://www.gatesfoundation.org/who-we-are/general-information/foundation-factsheet

24 V. Goel and N. Wingfield (2015). Mark Zuckerberg vows to donate 99% of his Facebook shares for charity. *New York Times*, 1 December.

25 Bezos Day One Fund (website). https://www.bezosdayonefund.org/

26 C. Clifford (2019). Billionaire Ray Dalio: 'Of course' rich people like me should pay more taxes. CNBC *Make It*, 8 April. Available at: https://www.cnbc.com/2019/04/08/bridgewaters-ray-dalio-of-course-rich-people-should-pay-more-taxes.html

27 PATH group (2015). The Meningitis Vaccine Project: a groundbreaking partnership. 13 June. https://www.path.org/articles/about-meningitis-vaccine-project/

28 D. Fluskey (2019). Why fewer people are giving to charity and what we can do about it. *Civil Society*,

8 May. Available at: https://www.civilsociety.co.uk/
voices/daniel-fluskey-why-fewer-people-are-giv-
ing-to-charity-and-what-we-can-do-about-it.html

29 State Street Global Advisers (2018). Global
Retirement Reality Report Ireland 2018: Ireland
Snapshot. State Street Corporation.

30 GiveWell (n.d.) Top charities. Available at: https://
www.givewell.org/charities/top-charities

31 F. Reddan (2016). Charities reveal how every €1
donated is spent. *Irish Times*. 2 January.

32 C. Mortimer (2015). One in five charities spends
less than half their total income on good causes,
says new report. *The Guardian*.

33 Charities Aid Foundation (2019). Charity Land-
scape Report 2019. Available at: https://www.cafon-
line.org/about-us/publications/2019-publications/
charity-landscape-2019

34 O.F. Williams (2016). *Sustainable Development: The
UN Millennium Development Goals, the UN Global
Compact, and the Common Good*, John W. Houck
Notre Dame Series in Business Ethics. University
of Notre Dame Press.

35 M.B. Weinberger (1987.) The relationship between
women's education and fertility: selected findings
from the world fertility surveys. *International
Family Planning Perspectives*, Vol. 13, 35–46.

36 S. Konrath (2017). Six reasons why people give
their money away, or not. *Psychology Today*, 26
November.

37 S. Konrath and F. Handy (2017). The development
and validation of the motives to donate scale.
Non-profit and Voluntary Sector Quarterly, 47:
347–375.

38 H. Cuccinello (2020). Jack Dorsey, Bill Gates and
at least 75 other billionaires donating to pandemic
relief. *Forbes*, 15 April. Available at https://www.
forbes.com/sites/hayleycuccinello/2020/04/15/jack-
dorsey-bill-gates-and-at-least-75-other-billion-
aires-donating-to-pandemic-relief/#6700456621bd

39 Charities Aid Foundation (2019). CAF UK Giving
2019. Available at: https://www.cafonline.org/docs/
default-source/about-us-publications/caf-uk-giv-
ing-2019-report-an-overview-of-charitable-giving-
in-the-uk.pdf

CHAPTER 13

1 Adventures in Energy (n.d.). How are oil and
natural gas formed? Available at: http://www.
adventuresinenergy.org/What-are-Oil-and-Natu-
ral-Gas/How-Are-Oil-Natural-Gas-Formed.html

2 M. Vassiliou (2018). *Historical Dictionary of the
Petroleum Industry* (2nd edition). Rowman &
Littlefield.

3 Wikipedia (n.d.) List of countries by proven oil
reserves. Available at: https://en.wikipedia.org/
wiki/List_of_countries_by_proven_oil_reserves

4 G. Liu (ed.) (2012). *Greenhouse Gases*. IntechOpen.

5 R. Jackson (2018). *The Ascent of John Tyndall: Victorian Scientist, Mountaineer and Public Intellectual.*

Oxford University Press.

6 S. Arrhenius (1896). On the influence of carbonic
acid in the air upon the temperature of the ground.
*London, Edinburgh, and Dublin Philosophical
Magazine and Journal of Science*, 41:251, 237–276.

7 Z. Hausfather (2017). Analysis: Why scientists
think 100% of global warming is due to humans.
Available at: https://www.carbonbrief.org/analysis-
why-scientists-think-100-of-global-warming-is-
due-to-humans

8 D. Lüthi, M. Le Floch, B. Bereiter, T. Blunier, J.M.
Barnola et al. (2008). High-resolution carbon dioxide concentration record 650,000–800,000 years
before present. *Nature*, 453: 379–382.

9 MuchAdoAboutClimate (2013). 4.5 billion years
of the earth's temperature. 3 August. Available
at: https://muchadoaboutclimate.wordpress.
com/2013/08/03/4-5-billion-years-of-the-earths-
temperature/

10 J. Watts (2019). 'No doubt left' about scientific
consensus on global warming, say experts. *The
Guardian*, 24 July. Available at: https://www.
theguardian.com/science/2019/jul/24/scientif-
ic-consensus-on-humans-causing-global-warm-
ing-passes-99

11 J. Hansen et al. (2006). Global temperature
change. *Proceedings of the National Academy of
Sciences of the United States of America*, 103(39):
14288–14293. doi:10.1073/pnas.0606291103

12 J. Mouginot et al. (2019). Forty-six years of Green-
land ice sheet mass balance from 1972 to 2018.
*Proceedings of the National Academy of Sciences of the
United States of America*, Vol. 116: 9239–9244.

13 C. Nunez (2019) Sea level rise, explained. *National
Geographic*. Available at: https://www.nationalgeo-
graphic.com/environment/global-warming/sea-lev-
el-rise/

14 V. Masson-Delmotte et al. (2018) IPCC Report
2018: Global Warming of 1.5°C. An IPCC Special
Report on the impacts of global warming of 1.5°C
above pre-industrial levels and related global
greenhouse gas emission pathways. Available
at: https://www.ipcc.ch/site/assets/uploads/
sites/2/2019/06/SR15_Full_Report_High_Res.pdf

15 M. Bevis et al. (2019). Accelerating changes in
ice mass within Greenland, and the ice sheet's
sensitivity to atmospheric forcing. *Proceedings of
the National Academy of Sciences of the United States
of America*. Vol. 116, 1934–1939.

16 P. Griffen (2017). CDP Carbon Majors Report
2017. Available at: https://b8f65cb373b1b7b-
15feb-c70d8ead6ced550b4d987d7c03fcdd1d.ssl.cf3.
rackcdn.com/cms/reports/documents

17 *The Economist* (2019) Leader: A warming world. 21
September.

18 H. Ritchie and M. Roser (2020). Renewable energy.
Our World in Data. Available at: https://ourworld-
indata.org/renewable-energy

19 D. Cross (2019). Engulfed in plastic: life is at risk

in the planet's oceans sustainability. *The Times*. 3 September.

20   Green Home (n.d.). Energy – carbon footprint. Available at: https://www.greenhome.ie/Energy/Carbon-Footprint

21   S. Wynes and K.A. Nicholas (2017). The climate mitigation gap: education and government recommendations miss the most effective individual actions *Environmental Research Letters* 12: 074024.

22   J. Chen et al. (2019). Methane emissions from the Munich Oktoberfest. *Atmospheric Chemistry and Physics Discussions*, https://doi.org/10.5194/acp-2019-709

23   E. Chenoweth and M. Stephan (2012). *Why Civil Resistance Works: The Strategic Logic of Nonviolent Conflict*, Columbia Studies in Terrorism and Irregular Warfare. Columbia University Press.

CHAPTER 14

1   E.J. Emanuel et al. (2016). Attitudes and practices of euthanasia and physician-assisted suicide in the United States, Canada and Europe. *JAMA* 316(1), 79–90.

2   S. Andrew (2020). Where is euthanasia legal? Three terminally ill minors choose to die in Belgium, new report finds. *Newsweek*, 18 January.

3   Irish Hospice Foundation (n.d.). Study Session 3: Healthcare Decision-making and the Role of Rights. Available at: http://hospicefoundation.ie/wp-content/uploads/2013/06/Module_3.pdf

4   D. McDonald (2013). Marie Fleming loses Supreme Court 'Right-to-die' case. *Independent*, 29 April.

5   P. Hosford (2015). Gail O'Rourke found not guilty of assisting the suicide of her friend. *The Journal*, 28 April. Available at: https://www.thejournal.ie/assisted-suicide-a-crime-if-suicide-isnt-2073607-Apr2015/

6   Health Service Executive (n.d.). Euthanasia and assisted suicide. Available at: https://www.hse.ie/eng/health/az/e/euthanasia-and-assisted-suicide/alternatives-to-euthanasia-and-assisted-suicide.html

7   I. Dowbiggin (2005). *A Concise History of Euthanasia: Life, Death, God, and Medicine*, Critical Issues in World and International History. Rowman & Littlefield.

8   J. Lelyveld (1986). 1936 secret is out: doctor sped George V's death. *New York Times*, 28 September.

9   ProCon.org (2019). History of euthanasia and physician-assisted suicide. Available at: https://euthanasia.procon.org/historical-timeline/

10   A.M.M. Eggermont et al. (2018). Combination immunotherapy development in melanoma. *American Society of Clinical Oncology Education Book*, 38: 197–207.

11   BBC (n.d.) Ethics of euthanasia – introduction. Available at: http://www.bbc.co.uk/ethics/euthanasia/overview/introduction.shtml

12   J.A. Colbert et al. (2013). Physician-assisted suicide – polling results. *New England Journal of Medicine*, 369: e15.

13   J. Wood and J. McCarthy (2017). Majority of Americans remain supportive of euthanasia. Gallup. Available at: https://news.gallup.com/poll/211928/majority-americans-remain-supportive-euthanasia.aspx

14   O. Bowcott (2019). Legalise assisted dying for terminally ill, say 90% of people in UK. *The Guardian*. 3 March. Available at: https://www.theguardian.com/society/2019/mar/03/legalise-assisted-dying-for-terminally-ill-say-90-per-cent-of-people-in-uk

15   Populus (2019). Largest ever poll on assisted dying conducted by Populus finds increase in support to 84% of the public. Available at: https://www.populus.co.uk/insights/2019/04/largest-ever-poll-on-assisted-dying-conducted-by-populus-finds-increase-in-support-to-84-of-the-public/

16   H. Halpin (2019). 3 in 5 people in Ireland support the legalisation of euthanasia. *The Journal*, 1 December. Available at: https://www.thejournal.ie/legalisation-euthanasia-ireland-poll-4913894-Dec2019/

17   J. Pereira (2011). Legalising euthanasia or assisted suicide: the illusion of safeguards and controls. *Current Oncology*, 18, e38–e45.

18   M. Erdek (2015). Pain medicine and palliative care as an alternative to euthanasia in end-of-life cancer care. *Linacre Quarterly*, 82(2): 128–134.

19   S. Dierickx et al. (2018). Drugs used for euthanasia: a repeated population-based mortality follow-back study in Flanders, Belgium, 1998–2013. *Journal of Pain and Symptom Management*, 56: 551–559.

20   R.W. Olsen (1986.) Barbiturate and benzodiazepine modulation of GABA receptor binding and function. *Life Sciences* 39: 1969–76.

21   A. von Baeyer (1864). Untersuchungen über die Harnsäuregruppe. *Annalen*, 130:129.

22   C. de Bellaigue (2019). Death on demand: has euthanasia gone too far? *The Guardian*, 18 January. Available at: https://www.theguardian.com/news/2019/jan/18/death-on-demand-has-euthanasia-gone-too-far-netherlands-assisted-dying.

23   C. de Lore (2019). The Dutch ethics professor who changed his mind on euthanasia. *New Zealand Listener*,19 October.

24   S. Caldwell (2018). Dutch euthanasia regulator quits over dementia killings. *Catholic Herald*, 23 January.

25   S. Boztas (2018). Dutch doctor reprimanded for 'asking family to hold down euthanasia patient'. *The Telegraph*, 25 July.

26   G. Blobel (2013). Christian de Duve (1917–2013): biologist who won a Nobel Prize for insights into cell structure. *Nature*, 498: 300.

27   *Le Soir* (Belgium) (2013). Christian de Duve a choisi le moment de sa mort. Available at: https://www.lesoir.be/art/237537/article/actualite/belgique/2013-05-06/christian-duve-choisi-moment-sa-mort

CHAPTER 15

1   M. Blitz (2018). We already know how to build a time machine. *Popular Mechanics.*

2   J. Knight (2018). France has a brand new Defense Innovation Agency. *Open Organization.* Available at: https://open-organization.com/en/2018/11/03/francais-la-france-a-son-agence-de-linnovation-de-defense/

3   C. Fernández (2019). New Arup report reveals best and worst scenarios for the future of our planet. Arup. https://www.arup.com/news-and-events/new-arup-report-reveals-best-and-worst-scenarios-for-the-future-of-our-planet

4   M. Venables (2013). Why Captain Kirk's call sparked a future tech revolution. *Forbes*, 3 April.

5   A. Boyle (2017). Make it so, Alexa: Amazon adds a few new Star Trek skills to AI assistant's repertoire. *GeekWire.* 21 September. Available at: https://www.geekwire.com/2017/make-alexa-amazon-adds-star-trek-skills-ai-assistants-repertoire/

6   J. Merkoski (2013). *Burning the Page: The eBook Revolution and the Future of Reading.* Sourcebooks Inc.

7   C. Peretti (2020). Instagram CEO Adam Mosseri says an episode of 'Black Mirror' inspired the decision to test hiding likes. *Business Insider*, 18 January.

8   D. Mosher (2018). The US military released a study on warp drives and faster-than-light travel. Here's what a theoretical physicist thinks of it. *IFLScience!* Available at: https://www.iflscience.com/physics/the-us-military-released-a-study-on-warp-drives-and-faster-than-light-travel-heres-what-a-theoretical-physicist-thinks-of-it/

9   NASA (n.d.) Explore Moon to Mars. Available at: https://www.nasa.gov/topics/moon-to-mars/lunar-gateway

10   K. O'Sullivan (2019). Aviation emissions set to grow sevenfold over 30 years, experts warn. *Irish Times.* 26 January.

11   T. Gabriel (2017). Ryanair crisis: aviation industry expert warns 600,000 new pilots needed in next 20 years. *The Conversation*, 28 September. Available at: https://theconversation.com/ryanair-crisis-aviation-industry-expert-warns-600-000-new-pilots-needed-in-next-20-years-84852

12   J. Stewart (2018). A better motor is the first step towards electric planes. *Wired*, 27 September. Available at: https://www.wired.com/story/magnix-electric-plane-motor/

13   P. Birch (1982). Orbital ring systems and Jacob's ladders. *Journal of the British Interplanetary Society*, 35: 475–497.

14   J. Grant (2019). How will we travel the world in 2050? *Irish Examiner.* 1 September.

15   C. Loizos (2019). Boom wants to build a supersonic jet for mainstream passengers: here's its game plan. *TechCrunch*, 23 May. Available at: https://techcrunch.com/2019/05/22/boom-wants-to-build-a-supersonic-jet-for-mainstream-passengers-heres-its-game-plan/

16   C. Edwardes (2019). 'Spaceplane' that flies 25 times faster than the speed of sound passes crucial test. News.com.au, 10 April. Available at: https://www.news.com.au/technology/innovation/inventions/spaceplane-that-flies-25-times-faster-than-the-speed-of-sound-passes-crucial-test/news-story/97a98211666d58448981cf636f0dc619

17   E. Seedhouse (2016). *SpaceX's Dragon: America's Next Generation Spacecraft.* Springer International.

18   SpaceX (website). Starlink Mission. Available at: https://www.spacex.com/webcast

19   NASA (n.d.). Moonbase alpha overview. Available at: https://www.nasa.gov/offices/education/programs/national/ltp/games/moonbasealpha/mbalpha-landing-collection1-overview.html

20   BBC (2020). Yusaku Maezawa: Japanese billionaire seeks 'life partner' for Moon voyage. 13 Jaunary. Available at: https://www.bbc.com/news/world-asia-51086635

21   D. Palanker, A. Vankov and S. Baccus (2005). Design of a high-resolution optoelectronic retinal prosthesis. *Journal of Neural Engineering*, 2: S105–20.

22   C. Chang et al. (2019). Stable Immune Response Induced by Intradermal DNA Vaccination by a Novel Needleless Pyro-Drive Jet Injector. *AAPS PharmSciTech*, 21: 19.

23   S.M. McKinney, M. Sieniek and S. Shetty (2020). International evaluation of an AI system for breast cancer screening. *Nature*, 577: 89–94.

24   Phys.org (2009). Triage technology with a Star Trek twist. 27 May. Available at: https://phys.org/news/2009-05-triage-technology-star-trek.html

25   NASA (n.d.). National Space Biomedical Research Institute. Available at: https://www.nasa.gov/exploration/humanresearch/HRP_NASA/research_at_nasa_NSBRI.html

26   B. Curley (2017). Medical device used in 'Star Trek' is now a reality. *Healthline*, 11 August. Available at: https://www.healthline.com/health-news/medical-device-used-in-star-trek-is-now-a-reality#5

27   ClinicalTrials.gov (website). https://clinicaltrials.gov/

28   ClinicalTrials.gov (2020). Trends, charts and maps. Available at: https://clinicaltrials.gov/ct2/resources/trends

29   Microsoft (2013). IllumiRoom: peripheral project ed illusions for interactive experiences. Available at: https://www.microsoft.com/en-us/research/project/illumiroom-peripheral-projected-illusions-for-interactive-experiences/

30   J. Danaher and N. McArthur (2017). *Robot Sex: Social and Ethical Implications.* MIT Press.

31   K. Dowd (2019). Automation takes flight: a look at VC's soaring interest in robotics & drones. PitchBook. 25 March. Available at: https://pitchbook.com/news/articles/automation-takes-flight-a-lookat-vcs-soaring-interest-in-robotics-drones

32   C. Purtill (2019). Stop me if you've heard this one: a robot and a team of Irish scientists walk into a senior living home. Time, 4 October. Available at: https://time.com/longform/senior-care-robot/

33   J.D. Bernal (1929). *The World, the Flesh and the Devil: An Enquiry into the Future of the Three Enemies of the Rational Soul.* Kegan Paul, Trench, Trubner & Co.

34   FutureTimeline.net (n.d.) The 21st century. Available at: https://www.futuretimeline.net/21stcentury/21stcentury.htm

35   Bank My Cell (n.d.) How many smartphones are in the world? Available at: https://www.bankmycell.com/blog/how-many-phones-are-in-the-world

36   L. Ying (2019). 10 Twitter Statistics Every Marketer Should Know in 2020. Oberlo, 30 November. Available at: https://ie.oberlo.com/blog/twitter-statistics

37   J. D'Urso (2018). What's the least popular emoji on Twitter? *BBC Trending*, 29 July. Available at: https://www.bbc.com/news/blogs-trending-44952140

38   D. Noyes (2020) The top 20 valuable Facebook statistics. *Zephoria*, April. Available at: https://zephoria.com/top-15-valuable-facebook-statistics/